ELECTRICAL
INSTALLATION

Other titles from E & FN Spon

Building Regulations Explained
1992 Revision
J. Stephenson

CESMM3 Explained
B. Spain and L. Morley

Commercial Estimator
Marshall & Swift

Residential Estimator
Marshall & Swift

Construction Contracts
Law and Management
J. Murdoch and W. Hughes

Construction Tendering and
Estimating
J. I. W. Bentley

Effective Speaking
Communicating in speech
C. Turk

Effective Writing
Improving scientific, technical and
business communication
C. Turk and J. Kirkman

Estimating Checklists for Capital
Projects
2nd edition
The Association of Cost Engineers

Good Style
Writing for science and technology
J. Kirkman

Housing Defects Reference Manual
The Building Research Establishment
Defect Action Sheets
Building Research Establishment

Post-Construction Liability and
Insurance
Edited by J. Knocke

The Presentation and Settlement of
Contractors' Claims
G. Trickey

Project Budgeting for Buildings
D. Parker and A. Dell'Isola

Project Management Demystified
Today's tools and techniques
G. Reiss

Spon's Budget Estimating Handbook
2nd edition
Spain and Partners

Spon's Building Costs Guide for
Educational Premises
Tweeds

Spon's Construction Cost and Prices
Indices Handbook
M. Flemming and B. Tysoe

Spon's European Construction Costs
Handbook
Davis Langdon & Everest

Standard Method of Specifying for
Minor Works
3rd edition
L. Gardiner

Understanding JCT Building Contracts
3rd edition
D. M. Chappell

Write in Style
A guide to good English
R. Palmer

For more information on these and other titles please contact:
The Promotion Department, E & FN Spon,
2–6 Boundary Row, London, SE1 8HN.
Telephone 071-522 9966

SPON'S
CONTRACTORS' HANDBOOK

ELECTRICAL INSTALLATION

1995

Tweeds

CHARTERED QUANTITY SURVEYORS,
COST ENGINEERS, CONSTRUCTION ECONOMISTS

E & FN SPON
An Imprint of Chapman & Hall

London · Glasgow · Weinheim · New York · Tokyo · Melbourne · Madras

**Published by E & FN Spon, an imprint of Chapman & Hall,
2–6 Boundary Row, London SE1 8HN**

Chapman & Hall, 2–6 Boundary Row, London SE1 8HN, UK

Blackie Academic & Professional, Wester Cleddens Road, Bishopbriggs, Glasgow G64 2NZ, UK

Chapman & Hall GmbH, Pappelallee 3, 69469 Weinheim, Germany

Chapman & Hall USA., One Penn Plaza, 41st Floor, New York, NY 10119, USA

Chapman & Hall Japan, ITP-Japan, Kyowa Building, 3F, 2-2-1 Hirakawacho, Chiyoda-ku, Tokyo 102, Japan

Chapman & Hall Australia, Thomas Nelson Australia, 102 Dodds Street, South Melbourne, Victoria 3205, Australia

Chapman & Hall India, R. Seshadri, 32 Second Main Road, CIT East, Madras 600 035, India

First edition 1988
Fifth edition 1994

© 1994 E & FN Spon

Printed in Great Britain by St Edmundsbury Press, Bury St Edmunds, Suffolk

ISBN 0 419 18600 X

A catalogue record for this book is available from the British Library

∞ Printed on permanent acid-free text paper, manufactured in accordance with ANSI/NISO Z39.48-1992 and ANSI/NISO Z39.48-1984 (Permanence of Paper)

CONTENTS

CONTENTS

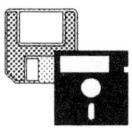

The Electrical Contractors' Association

ESCA House · 34 Palace Court · London W2 4HY · Tel: 071 229 1266 · Fax: 071 221 7344

I am sure that all electrical contractors will find this publication of value in providing useful information to assist their estimating practices and pricing decisions.

In this respect *Spon's Contractors' Handbook — Electrical Installation* supports the work undertaken by the Electrical Contractors' Association to provide its members with current factual information on all matters relating to their business interests — commercial, contractual, legal, employee relations, safety, training, technical and marketing.

In addition, the Association's specialist qualified staff provide individual advice to members across these areas: over 2000 enquiries are handled each month.

The Association backs its members with guarantees that their work will comply with appropriate standards and will be completed in the event of the original contractor's liquidation.

The ECA seeks to encourage professionalism in electrical contracting businesses and this handbook is a valuable contribution to this enterprise. I commend it to all contracting companies whatever their size.

Yours sincerely

R.D. Taylor
President

The Association of Electrical Installation Engineers

DIRECTOR. H McK SIMPSON · SECRETARY. G W BRYAN JENNINGS

Preface

This is the fourth edition of Spon's Electrical Installation Contractors' Handbook; the book is intended to help the electrician who is on the point of becoming self-employed or has already taken the plunge and is now operating a small contracting business.

Well established electrical contractors can also benefit by using this book and, from the helpful and constructive criticisms received, it would appear they are among the most regular users of the wealth of information contained in the following pages.

Advice on business matters is contained in the first three chapters and is set out in an easy-to-read style with many examples. It is not intended that these chapters should replace the need for professional advice when the occasion warrants; they are meant to complement this need and hopefully save money in consultation fees.

Chapter 4 sets out some of the basic principles of estimating but the main thrust of the book lies in Chapter 5 - Rates for Measured Work. The information contained in this chapter is intended as a base on which an electrical estimator can produce his quotations and tenders.

It cannot be stressed too strongly that, despite claims made on their behalf, price books should not be used as a literal source of information for preparing quotations, but as a base on which a contractor can formulate his own unique pricing data.

The measured items are presented in accordance with the requirements of the seventh edition of the Standard Method of Measurement of Building Works (SMM7): although this may be irrelevant to the needs of the smaller contractors it may well be required by those larger firms tendering for work on tender documents based upon SMM7.

It should be noted however that not all of the requirements of SMM 7 have been observed and it is hoped that the compromise will produce a balance which will suit the majority of readers.

There are many items affecting the value of electrical works. The speed and skill of individual workmen and the price paid for materials are factors which can directly affect the profitability of a job. The rates in Chapter 5 should be used as a basis only for preparing an estimate or quotation.

PREFACE

These comments do not detract from the real value of the information in this chapter. Over 4,000 item descriptions and unit rates are given which are based on 15,000 separate pieces of data providing a wealth of detailed cost information.

The original research for the first three editions of this book was carried out by Andy Williamson of Williamson Associates. This role has now been taken over by a team of quantity surveyors from Tweeds M & E section led by Tony Parry AMIEElecIE, MACostE assisted by Steven Bale. Similarly the original information on business matters was supplied by John Thirwell, but this task is now carried out by Grant Thornton, Chartered Accountants.

The editor has received a great deal of help from many sources in the research and the preparation of this book and would like to acknowledge the assistance given with grateful thanks.

Manufacturer	Equipment/Materials
Dimplex Heating Ltd	Electric heaters
Heatrae Sadia Ltd	Water heaters
MK Ltd	Telephone outlets Television outlets Data outlets General power/lighting Fittings and accessories PVC mini-trunking PVC conduit
Philips	Telephone and answering machines
Brother	Fax machines
Electric Services	Telecommunication, television and data cables and accessories
Static Systems Group	Staff paging and location
Power Centre Holdings Ltd	Steel trunking Plastic trunking and conduit
Gents Ltd	Clocks, fire detection and alarm equipment

Securitec Ltd	Intruder alarms, domestic
Channel Safety Systems Ltd	Security detection and alarm
W J Furse & Co Ltd	Earthing and bonding
Vent-Axia Ltd	Ventilation fans
Walsall Conduits Ltd	Steel conduit and trunking
Swifts of Scarborough Ltd	Cable tray
Davis Group Ltd	Cable ladder and support systems
BICC Cables Ltd	Cables
BICC Pyrotenax Ltd	MICC cables
Ottermill Ltd	Switchgear and distribution equipment
BILL Switchgear Ltd	Switchgear and distribution equipment
Dorman Smith Switchgear Ltd	Switchgear and distribution equipment
Simmonds Bros	Transformers
Philips Lighting Ltd	Luminaires and lamps
Friedland Ltd	Bells and chimes
Superswitch	Timers and controls
Backer	Test equipment
TMK	Test equipment
Paul Spain	Presentation

PREFACE

Edmundson Electrical Ltd helped by providing a schedule of discounts for some of the standard equipment and materials. It should be noted, however, that the discounts stated at the front of each section of this book should be used as a guide only because they could vary depending upon the size of the order and the relationship between the manufacturer/supplier and the contractor.

The editors would welcome constructive criticism of the book together with suggestions for improving its scope and contents. Whilst every effort has been made to ensure the accuracy of the information given in this publication, neither the editor nor the publishers accept liability in any way or of any kind resulting from the use made by any person of such information.

There are now many women working in the construction industry; the pronoun 'he' used throughout this book applies to both men and women.

Bryan J.D. Spain FInstCes, MACostE
TWEEDS (incorporating SPAIN AND PARTNERS)
Consulting Quantity Surveyors
Cavern Walks
8 Matthew Street
Liverpool L2 6RE

SMM6/SMM7 INDEX

The table set out below shows the link between the item classification of the requirements of SMM6 and SMM7.

SMM6	SMM7
A GENERAL RULES	
B ELECTRICAL INSTALLATIONS	V ELECTRICAL SUPPLY/POWER LIGHTING SYSTEMS
	W COMMUNICATIONS/SECURITY/CONTROL SYSTEMS
	Y MECHANICAL AND ELECTRICAL SERVICES MEASUREMENT

Classification of work

a. incoming services	V11 HV supply/distribution/public utility supply
	V12 LV supply/public utility supply
b. standby equipment	V32 Uninterrupted power supply
	V40 Emergency lighting
c. mains installation excluding final sub-circuits	V20 LV distribution
d. power installation	V22 General LV power
e. lighting installation	V21 General lighting
	V41 Street/area/flood lighting
	V42 Studio/auditorium/arena lighting
	V90 General lighting and power (small scale)
	W21 Projection
	W22 Advertising display

SMM6	SMM7
f. electric heating installation	V50 Electric underfloor heating V51 Local electric heating units
g. electric appliances	
h. electrical work associated with plumbing and mechanical engineering installations	Y53 Control components - mechanical
j. telephone installations	W10 Telecommunications
k. clock installation	W23 Clocks
l. sound distribution installation	W11 Staff paging/location W12 Public address/sound amplification W13 Centralized dictation
m. alarm system installation	W41 Security, detection and alarm W50 Fire detection and alarm
n. earthing system installation	W51 Earthing and bonding
p. lightning protection installation	W52 Lightning protection
q. special services	V30 Extra low voltage supply V31 DC supply W20 Radio/TV/CCTV W30 Data transmission W40 Access control W53 Electromagnetic screening W60 Monitoring W61 Central control W62 Building automation

SMM6/SMM7 INDEX

SMM6	SMM7
Equipment and control gear	V10 Electricity generation plant
	Y70 HV switchgear
	Y71 LV switchgear and distribution boards
	Y72 Contactors and starters
	Y92 Motor drives - electric
Fittings and accessories	Y73 Luminaires and lamps
	Y74 Accessories for electric services
Conduit, trunking and cable trays	Y60 Conduit and cable trunking
	Y62 Busbar trunking
Cables/final sub-circuits	Y61 HV/LV cables and wiring
Earthing	Y80 Earthing and bonding components
Ancillaries	Y82 Identification - electrical
	Y89 Sundry common electrical items
Sundries	Y81 Testing and commissioning of electrical services
Builders' work	P31 Holes/chases/covers/supports for services
Protection	A42 Contractor's services and facilities

It can be seen, however, that to follow the SMM7 layout exactly would produce a great deal of unnecessary repetition in this book: e.g. the items covered by section V20 would need to be repeated in V21, V41, V42, V90, and V22.

The technical contents, therefore, have been set out generally in accordance with SMM7 but rationalized in a concise and logical format as follows:

Introduction

There are two main ways that contractors prepare tenders and quotations. The first and probably the less frequent method is to insert rates against item descriptions in a tender document prepared by the prospective employer's professional advisers. This would occur on major schemes where a bill of quantities and/or a schedule of rates has been prepared. In this case the contractor would be able to examine and use the rates contained in Chapter 5 of this book as a basis for his bid. Usually, however, the contractor is either handed a plan and specification or merely invited to inspect the premises and prepare his offer without the benefit of any paperwork at all! In either of these two cases the contractor must take off his own quantities. Once he has done that, the contents of Chapter 5 can be used as a base on which he can prepare his quotation.

Whichever method is used the main value of this book lies in a sensible application of the thousands of rates in Chapter 5. People who have not used price books before sometimes query their value in that every craftsman does not work at the same speed, wide variations in discounts for materials can be obtained, each job has different cost-related circumstances, etc.

The answer to these criticisms is that successful users of price books are fully aware of these difficulties but overcome them by understanding the relationship between their own production costs and material discounts and those assumed in the book.

For example, a regular user of this book will know that the published rates may be a certain percentage higher or lower than his own costs. With this knowledge he can quickly prepare his quotations by using the book rates and making the appropriate percentage adjustment at the end.

The editors regularly receive comments from contractors using these books and it seems that most firms' costs fall in a band between 10% higher and lower of those in this book. Some firms have been good enough to state that they have been able to win contracts in competition by using the book's rates verbatim and have secured good profits as well.

INTRODUCTION

Careful thought must be given to the unique circumstances of each job undertaken and the percentage allowances for 'project labour factors' at the beginning of Chapter 5 should be studied carefully and their potential cost in all quotations.

Profits can turn into losses without an assessment of the effect that each project's elements can have on the rates and the editors urge readers to consider this matter very carefully.

Someone once said that price books were like guns - dangerous in the wrong hands! There is some truth in this but using the information wisely can save the one commodity you cannot buy - time.

And that in itself should be of great value to a busy contractor.

Chapter 1
Starting up in business

Before committing himself by giving up his job, the would-be businessman should consider carefully whether he has the skills and also the temperament to survive in the highly competitive self-employed market. He should also do a lot of research and seek out as much information as possible about how to run a business. He should then know whether his business idea is likely to work in practice and have some idea of the new, strange and sometimes complex requirements of running such a business.

INITIAL RESEARCH

Matters to be researched should include the following.

Finance

Assess what funds will be needed and when will items such as premises, plant, transport, tools, initial stocks of materials, wages, overheads and the proprietors' living expenses need to be paid for before the cash from work done begins to flow in. How is the price of work done to be calculated?

All of this needs a proper 'business plan' and if the bank or other sources of finance are to be tapped such a plan is essential. Fortunately help in its preparation is available from a number of agencies sponsored by goverment, local authorities and industry, usually at little or no cost.

Testing the market

Talk to as many traders already operating in the same field as possible. Try to identify if the need is in the industrial, commercial, local government or domestic field. Talk to likely customers and clients and consider whether it is possible to improve on what they are being offered at present in terms of price, quality, speed, convenience, follow-up, etc.

Advertising

Those entering the domestic side of the business will need to think about the best way to reach potential customers. Are local word-of-mouth recommendations enough to provide reasonably continuous work? If not, what would be the most effective method?

Advertising is costly and it is a waste of funds to place an advert in a paper which circulates in areas A, B, C and D if the business can only cover area A. Advice on the best medium and the content of adverts is available at the Small Business Service.

Experience and training

After making an objective appraisal of the likely market for the goods or services, then the extent to which one is equipped to satisfy its requirements should be considered. Gaps in experience might be filled by a change of present duties or of employer, whilst some lack of skills can probably be overcome by taking a training course.

INITIAL INFORMATION

Training and Enterprise Councils (TECs)

There is no shortage of information about the many aspects of starting and running your own business: finance; marketing; legal requirements; developing your business idea; taxation, etc., are all the subject of a mountain of books, pamphlets, guides and courses. Indeed the likelihood is that the aspiring businessman will be overloaded and thoroughly confused rather than left high and dry without guidance. Nor is it necessary to pay out a lot of money.

A good place to start for both information and advice is your local TEC. This comprises a board of directors drawn from the top men in local industry, commerce, education, trade unions etc., who, together with their staff and experienced business counsellors, assist both new and established concerns in all aspects of running a business. This takes the form not only of across-the-table advice but also, if desired, hands-on assistance in management, marketing, finance, etc. There are also training courses and seminars available in most areas.

Contact can be made through the local job centre, Citizens' Advice Bureau or by ringing Freephone Enterprise 0800 222999.

INITIAL INFORMATION

Business links

There are organizations currently being established with a view to providing a 'one-stop-shop' for advice and assistance to owner-managed businesses. When established they will often replace the need to contact TECs and many of the other official organizations listed below.

Enterprise Agencies

Another useful source of free help and advice is the local Enterprise Agency. Run by local businesses for small and developing concerns it covers similar ground to the TEC.

The address and telephone number of the agency in your area can be obtained by ringing 'Business in the Community' on 071-253 3716.

The Business Start-up Scheme

This is an allowance of £40 a week, in addition to any income made from your business, paid for forty weeks.

To qualify you must be at least 18 and under 65, work at least 36 hours per week in the business and have been unemployed for at least six weeks or fall into one of the other categories - disabled, ex-HMF, redundant etc.

The first step is to get the booklet on the subject from your local job centre or TEC; all the details are in it, including how and where to apply. Once in receipt of the Enterprise Allowance you will also have the benefit of advice and assistance from an experienced businessman from your TEC. All the initial counselling services and training courses are free.

Potential customers and trade contacts

Many who become self-employed in the construction industry already have experience as employees. Use these contacts to check the market, establish the sort of work which is available and the current contract rates. In the domestic market check on the competition for prices, standards of work and service provided, customer complaints and types of advertising.

Try to get firm promises of work before the start-up date.

Potential suppliers

Canvass local suppliers for the best prices, credit terms, minimum order, discounts offered and delivery times. Remember cash, is the lifeblood of business and a supplier who gives 30 days' credit and delivers small quantities at 24/48 hours' notice may be a better buy than one with lower prices but who operates on cash and carry terms only. It is important not to overstock, but carry only what is needed for current requirements. Do not tie up scarce and expensive money in stock which may not be used for weeks or even months.

Banks

Approach banks for information about a business current account and financial services available. Find out about the types of loan required; from an overdraft for working capital to medium and long terms loans including the Government's Loan Guarantee Scheme (see page 10) for the purchase of plant and machinery and alterations to premises. See the TEC first for information about the best approach and find out what the bank manager will need. Shop around several banks and branches if you are not satisfied at first; managers vary widely in their views of what is a viable business proposition. Most banks have useful free information packs to help business start-up.

Point of contact: the local bank manager.

HM Inspector of Taxes

Make a preliminary visit to the local tax office enquiry counter for their publications:

IR 14/15	*Construction Industry Tax Deduction Scheme*
IR28	*Starting in Business*, and, if needed,
IR40	*Conditions for Getting a Sub-contractor's Tax Certificate*
IR53	*PAYE for Employers* (if you employ someone)
IR56/N139	*Employed on Self-employed.*

The onus is on the taxpayer to notify the Inland Revenue that he is in business. Failure to do so may result in the imposition of interest and penalties. Either send a letter or use the form provided in the middle of the *Starting in Business* booklet.

Point of contact: telephone directory for address.

VAT Office

Registration for VAT is required if:

1. At the end of any month the value of taxable supplies in the past twelve months has exceeded the annual threshold;

2. There are reasonable grounds for believing that the value of taxable supplies in the next 30 days will exceed the annual threshold.

Taxable supplies include any zero-rated items. From 1 December 1993 the annual threshold is £45,000. Failure to register is an offence punishable by the imposition of financial penalties.

The VAT office also carries a number of useful publications including:

700	*The VAT Guide*
700/1	*Should I be Registered for VAT?*
700/12	*Filling in Your VAT Return*
700/21	*Keeping Records and Accounts*
708/2	*Application of VAT to the Construction Industry*
731	*Cash Accounting*
732	*Annual Accounting*
742	*Land and Property*

Notes on the application of VAT to the Construction Industry are in Chapter 4 together with information about the new 'Cash accounting scheme' and the introduction of annual VAT returns.

Point of contact: telephone directory for address.

DSS Office

Class 2 contributions payable by the self-employed may be paid either in cash or by direct debit through a bank account. Call at the local office to make the necessary arrangements. Class 4 contributions which are also payable by the self-employed are collected along with the income tax by the Inland Revenue and no special action by the businessman is required.

Ask at the DSS office for the following publications:

NI 41	*NI Guide for the Self-employed*
NI 27A	*People with Small Earnings from Self-employment*
NI 35	*NI for Company Directors*
NI 225	*Direct Debit - The Easy Way to Pay,* and for employers
NP 15	*Employers' Guide to NIC*
NI 227	*Employers' Guide to Statutory Sick Pay*

Point of contact: telephone directory for address.

Local authorities

Authorities vary in the provisions made for small businesses but all have been asked to simplify and cut delays in planning applications. In Assisted Areas and Enterprise Zones rent-free periods and reductions in rates may be available on certain industrial and commercial properties. As a preliminary to either purchasing or renting business premises the following booklets will be very helpful. *A Step by Step Guide to Planning Permission for Small Businesses* and *Business Leases and Security of Tenure* are both issued by the Department of Employment and are available at council offices, Citizens' Advice Bureaux and TEC offices.

Some authorities run training schemes in conjunction with local industry and educational establishments.

Point of contact: usually the planning department - ask for the Industrial Development or Economic Development Officer.

Department of Trade and Industry

The current package of assistance from this department is called the Enterprise Initiative and is geared more to the needs of existing businesses than new start-ups. It ranges widely over all aspects; marketing, design, quality, management, finance, etc., and consists essentially of assessing the requirements of the business. The counsellor will keep an eye out for untapped resources, inefficient work systems and unrealized potential and will recommend specialist consultants to come in and advise. The department would pay all costs of the initial survey and one-half (or two-thirds in development areas) of the costs of any consultancy between 5 and 15 man days.

Point of contact: phone 071-215 5000 and ask for the address and phone number of the nearest DTI office and copies of their explanatory booklets.

Department of the Environment

From 1 April 1992 new regulations are in force under the Environmental Protection Act 1990 relating to all forms of waste other than normal household rubbish. Any concern which produces, stores, treats, processes, transports, recycles or disposes of such waste has a 'duty of care' to ensure it is properly discarded and dealt with. Practical guidance on how to comply with the law (it is a criminal offence punishable by a fine not to) is contained in a booklet *Waste Management*: *The Duty of Care: A Code of Practice*, obtainable from HMSO Publications Centre, PO Box 276, London SW8 5DT. Phone 071-873 9090. Price £5.00.

INITIAL INFORMATION

Accountant

The services of an accountant are to be strongly recommended from the beginning because the legal and taxation requirements start immediately and must be properly complied with if trouble is to be avoided later. A qualified accountant must be used if a limited company is being formed and for all types of business the accountant should be able to give advice on a whole range of business issues from, for example, book-keeping to grant aid, from tax planning and compliance to finance raising and will clearly help in preparing annual accounts.

It is worth spending some time finding an accountant who has other clients in the same line of business and is able to give sound advice particularly on taxation and business finance and is not so overworked that damaging delays in producing accounts are likely to arise. Ask other traders whether they can recommend their own accountant. Visit more than one firm of accountants, ask about the fees they charge and how much the production of annual accounts and agreement with the Inland Revenue are likely to cost and how long the work will take. A good accountant is worth every penny of his fees but do not hesitate to challenge him if his service is unsatisfactory.

Solicitor

Many businesses operate without the services of a solicitor but there are a number of occasions when legal advice should be sought. In particular no-one should sign the lease of premises without asking a solicitor what they are committing themselves to because it is not unusual for a business to be put into financial difficulty through unnoticed liabilities in its lease. Either an accountant or solicitor will help with drawing up a partnership agreement which all partnerships should have.

A solicitor will also help to explain complex contract terms and prepare draft contracts if the type of business being entered into requires them.

Insurance broker

Policies are available to cover many aspects of business including:

employer's liability - compulsory if the business has employees;
public liability - essential in the construction industry;
motor vehicles;
theft of stock, plant, money, etc.;
fire and storm damage;
personal accident and loss of profits.

Brokers are independent advisers who will obtain competitive quotations on your behalf. See more than one broker before making a decision - their advice is normally given free and without obligation.

Point of contact: telephone directory or write for a list of local members to:

> The British Insurance Brokers Association
> Consumer Relations Department
> BIBA House
> 14 Bevis Marks
> London EC3A 7NT (phone: 071-623 9043)

or contact

> The Association of British Insurers
> 51 Gresham Street
> London EC2V 7HQ (phone: 071-600 3333)

who will supply free, a package of very useful advice files specially designed for the small business.

The Health and Safety Executive

The Executive operates the legislation covering everyone engaged in work activities and the following free literature is available:

HSE 16   *The Law on Health and Safety at Work*
IND(G)14(L) *Compliance with Health and Safety Legislation at Work*
HSE4   *Short Guide to Employer's Liability (Compulsory Insurance) Act 1969*

The Executive has issued a very useful set of 'Construction Health Hazard Information Sheets' covering such topics as handling cement, lead and solvents, safety in the use of ladders, scaffolding, hoists, cranes, flammable liquids, asbestos, roofs and compressed gases, etc. A pack of these may be obtained free from your local HSE office or:

> The Health & Safety Executive
> St. Hugh's House
> Stanley Precinct
> Bootle
> Merseyside L20 3QY (phone: 051-951 4381)

FINANCE

Working out how much will be needed

The businessman should estimate in advance what funds will be needed to start up and to run the business for at least the first 12 months. If the forecast is to be reasonably accurate he must make some early decisions about:

1. The premises where the business will be based, what initial repairs, alterations, etc., to them are required and what will be the total cost.

2. What plant, equipment and transport is needed, whether it is to be leased or purchased and again what the cost will be.

3. How much stock of materials, if any, should be carried (the bare minimum only should be acquired so a reliable supplier should be sought out).

4. What will be the weekly bill for overheads, wages and the proprietor's living costs and how these are to be met until the cash for work done starts to flow in.

5. What type of work is going to be undertaken, how much profit margin can realistically be obtained and how often invoices are to be presented.

The above are just some of the items that should be covered. If the proprietor is fortunate enough to have some capital, or some is available within the family, he might get by without doing more than a rough estimate based on his initial research and knowledge of the trade. If however there is the need to seek finance from a bank or other financial institution a much more detailed statement will be required for which the help of an accountant or TEC adviser should be sought.

Sources of funds

Finance, like charity, often begins at home and the would-be businessman should make a realistic assessment of his net worth including the value of his house after deducting any mortgage(s) outstanding on it, his savings, any car or van owned and any sums which his family are prepared to contribute and deduct any private borrowing which will come due for payment in the next 24 months. The whole of these funds may not of course be available (for instance, money which has been loaned to a friend or relative who is known to be unable to repay at the present time). It may not be desirable that it should all be put at risk on a business venture. Establish therefore (a) how much cash it is proposed to invest in the business and (b) whether the family home will be made available for any business borrowings.

Whilst it may be wise not to pledge too much of the family assets, it has to be remembered that the bank will be looking closely at the degree to which the proprietor has committed himself to the venture and will not be impressed by an applicant for a loan who is prepared to risk only a small fraction of his own resources.

Having decided how much of his own funds to contribute, the businessman can now see the level of shortfall and consider how best to fill it. Consideration should be given to partners where the shortfall is large and particularly when there is a need for heavy investment in fixed assets such as premises and capital equipment. It may be worthwhile starting a limited company with others also subscribing capital and to allow the banks to take security against the book debts. The first outside source of money to which most businessmen turn is the bank and a book could be written solely on the do's and don'ts of approaching a bank manager. Here are a few tips:

1. Have a proper business plan to present to him including a cash flow forecast (12 months is usual), also an opening statement of affairs, projected profit and loss accounts and balance sheets for two years and a written statement describing the whole business venture. Use conservative estimates which tend to understate rather than overstate the forecast sales and profits.

2. Know the figures in detail - and don't leave it to an accountant to explain them for you. The bank manager is interested in the businessman not his advisers and will be impressed if he has a sound grasp of the financing of his business.

3. Understand the difference between short and long term borrowing; know how much is needed of each and be ready to explain how the business will be able to repay the bank its money.

4. Ask about the Government Loan Guarantee Scheme if there is a shortage of security for loans. Under this scheme the Government guarantees 70% of the loan up to £100,000 (now up to 85% for loans to businesses in the Inner City Task Force Areas) but if you have been in business for more than two years the limit is £250,000. In return there is an insurance of 0.5% on fixed rate interest arrangements and 1.5% on variable bank interest arrangements. Repayment is over a period from 2 to 7 years.

There are a number of other financial institutions in the 'venture capital' market that can help well established businesses, usually limited companies, wishing to expand and also for some well conceived start-ups. They will provide a flexible package of equity and loan capital in the range of £25,000 to £1,000,000. Usually the deal entails the institution having a minority interest in the voting share capital and a seat on the board of the company. Arrangements for the eventual purchase of the shares held by the finance company by the private shareholders are also normally incorporated in the scheme.

Contact points for information and advice are bank managers, accountants and the TEC. If the outside investor in the business is an individual he will probably wish to invest within the terms of the Enterprise Investment Scheme which enables him to get tax relief at 20% on the amount of his investment. The rules are complex and professional advice is essential.

The Royal Jubilee and Prince's Trusts

These trusts, through the Youth Business Initiative, provide bursaries of not more than £1,000 per individual to selected applicants who are unemployed and aged 25 or under. Grants may be used for tools and equipment, transport, fees, insurance, instruction and training, but not for working capital, rent and rates, new materials or stock. They operate through a local representative whose name and address may be ascertained by contacting The Prince's Youth Business Trust, 5 Cleveland Place, London SW1Y 6JS: Phone 081-968 3713.

THE CONSTRUCTION INDUSTRY TAX DEDUCTION SCHEME

General

The Construction Industry Tax Deduction Scheme is known universally as the '714' Scheme, after the number of the official form around which the whole system revolves. The government is proposing important changes to the scheme but they are still under discussion.

The businessman should visit his local income tax enquiry office and obtain copies of the Revenue booklet IR 14/15 and leaflet IR 40 which explain the conditions under which the Revenue will issue a 714 certificate.

The scheme operates whenever a 'contractor' makes a payment to a 'sub-contractor'.

If the sub-contractor does not hold a valid tax certificate (714I, 714P, 714C or 714S) issued to him by the Inland Revenue then the contractor *must* deduct 25% tax from the whole of any payment made to him (excluding the cost of any materials). If, however, he holds such a certificate the payment may be made in full without deducting tax (in the case of a 714S there is a weekly limit after which tax is deductible).

A business is not obliged by law to seek an exemption certificate and can legally work in the construction industry without one. As a matter of practice, however, many main contractors are reluctant to undertake the additional paperwork required when tax has to be deducted and accounted for to the Revenue and will give work to a sub-contractor with a 714 certificate in preference to one without.

A small business that does work only for the general public and small commercial concerns, is outside the scheme and does not need a 714 certificate to trade. If, however, it engages other contractors to do jobs for it, the business would have to register under the scheme as a contractor and deduct tax from any payments made to a sub-contractor who did not produce a valid 714 certificate.

Obtaining a 714 certificate

There is a special application form which may be obtained from tax offices. There are a number of conditions which have to be met before the Revenue will issue a certificate. In general terms these are:

1. The applicant must be working as a sub-contractor in the UK in the construction industry;

2. The business must be run from proper premises with the usual business facilities, and records must be kept and a business bank account operated;

3. The applicant must have been employed or self-employed in the UK for a continuous period of 3 years in the 6 years before the date of application; short breaks in employment will not be taken into account unless they exceed 6 months in total (but see 5);

4. The applicant must have a satisfactory tax and NIC record: (the Inland Revenue will check the NI contribution position with the DSS);

5. School leavers who can show that in the 6 years up to the date of application they were in full-time education or training for a continuous period of three years, or in full-time education or training for part of the three years and unemployed or in self-employment for the rest, may apply for a special certificate (714S);

6. Those who satisfy all the conditions except 3 may also obtain a special certificate (714S) if they can arrange for a bank to guarantee the tax payable on the amounts received in full (leaflet IR 40 sets out the conditions in details).

The 714 certificate

There are four types:

714I - which is issued to individuals;
714P - which is issued to partners;
714C - which is issued to most limited companies;
714S - which is issued as explained above.

The I, P and S certificates include a photograph of the individual to whom it is issued in addition to his name, NIC number, signature and business name (if any).
The certificate also has a serial number and an expiry date. Vouchers numbered 715 (in the case of a special certificate, 715S) are issued along with the 714I, P and S.

The 714 system in operation

Holders of certificates 714I and P (including company directors holding 714Ps)

Before paying a sub-contractor in full, without any deduction of tax, the contractor must carry out detailed checks on the 714 certificate (the original, not a copy) and the sub-contractor must produce the 714 for this purpose. On being paid the sub-contractor must complete a voucher 715 and give it to the contractor, showing, among other things, the amount of the payment received. The contractor must send the 715s to the Revenue monthly unless his monthly deduction plus any PAYE and NIC for employees total less than £450, when he may choose to pay quarterly.

Holders of 714C certificates

A sub-contractor which is a limited company may choose to produce to the contractor either the 714C itself or a 'certifying document'. Whichever method is used the documents have to be checked in detail by the contractor, if necessary by telephoning the company itself to confirm that the person presenting the document is their authorized representative.
If the contractor is satisfied he must pay the sub-contractor in full; 715s

are not used but both parties record the details of the payment in their business records.

Holders of 714S certificates

If the weekly payment to the sub-contractor (excluding the cost to him of any materials purchased directly from another person) does not exceed the limit shown on the front of the 715S voucher, then the procedure outlined above for a certificate 714I and P is followed and a voucher 715S is given to the contractor on receipt of the payment.

If, however, the payment exceeds the amount shown for the week then the 714I and P procedure is followed in respect of the amount shown and the procedure below is carried out for the balance of the payment.

Sub-contractors with no certificate and those with S certificate payments in excess of the weekly limit

The contractor is obliged to deduct tax from all payments (excluding the cost of directly purchased materials) and to account to the Revenue for all amounts so withheld. To enable the sub-contractor to prove to the inspector of taxes that he has suffered this tax deduction the contractor must give him a certificate on form SC60 showing the amount withheld. These SC60 forms must be carefully filed for production to the Inspector after the end of his accounting year along with his business profit and loss account and balance sheet. Any tax deducted in this way over and above the sub-contractor's proper, agreed liability for the year will be repaid by the Inland Revenue.

At the end of the day the sub-contractor will pay the same amount of income tax whether or not he has a certificate, but those without one will have suffered a severe restriction in their cash flow until the repayment is made.

The main contractor periodically has to send the 715 vouchers to the Inland Revenue Computer Centre in Liverpool and make an annual return to that Centre also. The date for filing the end-of-year return of payments to uncertificated sub-contractors has been extended by one month to 19 May. It all amounts to a heavy burden on the trader with penalties awaiting those who are sloppy, dilatory or dishonest in its operation.

Miscellaneous points

1. A payment includes anything paid out by the contractor such as a 'sub' or a loan, whether by cash, cheque or credit and whether direct to the sub-contractor or to his nominee.

2. The cost of subsistence and travelling expenses reimbursed by the contractor is included in the amount on which the tax deduction is calculated.

3. The scheme is policed by the Revenue in much the same way as they inspect PAYE documents, and records for the scheme have to be made available on request.

4. A contractor is liable to pay to the collector of taxes all amounts which he *should* have deducted from sub-contractors, whether he made the deductions or not. He may, however, be excused from having to pay if he can show he took reasonable care and made the error in good faith.

   If the deductions have not been properly made the sub-contractor himself will be asked to account for his own correct liability.

5. In a situation where tax *has* been deducted from payments made by him but the contractor has failed to account for it to the collector, the sub-contractor would be unlikely to recover any over-deductions from the Revenue.

6. Disputes between contractors and sub-contractors about the amount of any deductions should be referred to the inspector for a ruling.

Note

   The above is merely a summary of the very detailed instructions contained in IR 14/15 paragraphs 65-129 which must be carefully studied by anyone involved in operating the scheme either as a contractor or sub-contractor.

VAT

The general rule about liability to register for VAT is given in the VAT office notes above. It is possible to give here only a brief outline of how the tax works. The rules which apply to the construction industry are extremely complex and all traders must study *The VAT Guide* and other publications.

   The amount of tax to be paid is the difference between the VAT charged out to customers *(output tax)* and that suffered on payments made to suppliers for goods and services *(input tax)*. Unlike income tax there is no distinction in VAT for capital items so that the tax charged on the purchase of, for example, machinery, trucks and office furniture, will normally be reclaimable as Input Tax. One important exception to this is that the input tax on a car cannot be reclaimed even though it is used wholly for business purposes.

   VAT is payable in respect of 3-monthly periods known as 'tax periods' and you can apply to have the group of tax periods which fits in best with your financial year. The tax must be paid within one month of the end of each tax period. Traders who receive regular repayments of VAT can apply to have them monthly rather than quarterly.

Not all types of goods and services are taxed at 17.5% (i.e., at the standard rate), some are exempt and others are zero-rated.

Zero-rated

This means that no VAT is chargeable on the goods or services but a registered trader can reclaim any *input* tax suffered on his purchases. For instance a builder pays VAT on the materials he buys, but if he is constructing a new dwelling house, this is zero-rated and he may reclaim this VAT or set it off against any VAT due on other standard rated work.

Exempt

Supplies which are exempt are less favourably treated than those which are zero-rated. Again no VAT is chargeable on the goods or services but the trader cannot reclaim any *input* tax suffered on his purchases.

Standard-rated

All work which is not specifically stated to be zero-rated or exempt is standard-rated, i.e., VAT is chargeable at the current rate of 17½% and the trader may deduct any *input* tax suffered when he is making his return to the Customs and Excise.

If for any reason a trader makes a supply and fails to charge VAT when he should have done so (e.g. mistakenly assuming the supply to be zero-rated) he will have to account for the VAT himself out of the proceeds. If there is any doubt about the VAT position it is safer to assume the supply is standard rated, charge the appropriate amount of VAT on the invoice and argue about it later.

Time of supply

The *time* at which a supply of goods or services is treated as taking place is important and is called the 'tax point'. VAT must be accounted for to the Customs and Excise at the end of the accounting period in which this 'tax point' occurs. For the supply of *goods* which are 'built on site' the 'basic tax point' is the date the goods are made available for the customer's use, whilst for *services* it is normally the date when all the work except invoicing is completed.

However, if you issue a tax invoice or receive a payment *before* this 'basic tax point' then that date becomes the tax point.

In the case of contracts providing for stage and retention payments the tax point is either the date the tax invoice is issued or when payment is received, whichever is the earlier.

All the above requirements apply to sub-contractors and main contractors and it should be noted that, when a contractor deducts income tax from a payment to a sub-contractor (because he has no valid 714), VAT is payable on the full gross amount *before* taking off the Income Tax (see Chapter 3).

See below for examples of VAT payments and repayments.

Examples of how VAT works are as follows:

(a) Sales invoices total	£1000
plus VAT @ 17½%	<u>175</u>
The customer pays	<u>£1175</u>
Purchase invoices total	£600
plus VAT @ 17½%	<u>105</u>
Total paid	<u>£705</u>

In this example the *output* tax is £175 and the *input* tax is £105 which means that the trader owes £70 to the Customs and Excise.

If in the tax period machinery costing £2000 plus VAT £105 had also been purchased the figures would be:

(b) Sales invoices total	£1000
plus VAT @ 17½%	<u>175</u>
The customer pays	<u>£1175</u>

STARTING UP IN BUSINESS

Purchase invoices total	£2600
plus VAT @ 17½%	<u>455</u>
Total paid	<u>£3055</u>

In this example the *output* tax is also £175 but the *input* tax is £455 so that a refund of £280 is due to the trader from the Customs and Excise.

Chapter 2
Running the business

Many businesses are run without adequate information being available to check trends in their vital areas, e.g. marketing, money and managerial efficiency. It is vital to look critically at all aspects of the business to maximize profits and eliminate inefficiency.

Proprietors often have the feeling that the business should be 'doing better' than it is, without being able to identify what is going wrong. Sometimes there is the worrying phenomenon of a steadily increasing work programme coupled with a persistently reducing bank balance or rising overdraft.

Some useful ways of checking the position and identifying problem areas are given below.

Marketing

Whilst management and finance are concerned with the internal running of a business the market is where it makes contact with the outside world in the shape of its competitors and customers. Throughout his business life the entrepreneur should study carefully the methods and approach of the former and the needs and wishes of the latter. A shortcoming frequently found in ailing concerns is that the proprietor thinks he knows better than his customers what they want.

The term 'market research' sounds both difficult and expensive but a very simple form of it can be done quite effectively by the businessman and his sales staff. First, identify the type of person or business to whom the products or service are likely to appeal, finding out and recording what it is the customer wants in terms of price, quality, design, payment terms, follow-up service, guarantees, services.

The initial approach might be by leaflet or letter followed by a personal call. As an on-going part of management all staff with customer contact should be encouraged to enquire about and record customer preferences, complaints, etc., and feed it back to management.

Other sources of information are friends in the trade, business journals, trade exhibitions, suppliers, representatives, etc., from whom information about trends, new techniques and products can be obtained and studied.

Valuable information can also be gained from studying competitors and the following questions should be asked:

- what do they sell and at what prices?

- what inducements to buy do they offer their customers (e.g. credit facilities, guarantees, free offers, discounts, etc.)?

- how do they reach their customers (local/national advertising; mail shots; salesmen; local radio and TV)?

- what are the strongest aspects of their appeal to customers and have they any weaknesses?

The businessman should apply all the information gathered from customers and competitors to his own range of products with a view to making sure he is offering the right product at the right price in the most attractive way and in the most receptive market.

In a small business where the proprietor is also his own salesman he must give careful thought to how he can best present his product and himself. If he is working solely within the construction industry his main problems are likely to centre on getting a 714 and on exploiting as fully as possible trade contacts to get sub-contract work.

However, for those who serve the general public, presentation can be a vital element in getting work. The customer is looking for efficiency, reliability and honesty in a trader and quality, price and style in the product. To bring out these facets in discussion with a potential customer is a skilled task and for newcomers to business and those who have not had as much success to date as they hoped for, a short course on marketing techniques could pay handsome dividends.

The Enterprise unit of the Manpower Services Commission will give the names and addresses of such courses locally - contact it through the job centre or make an appointment to see a TEC adviser with marketing experience and talk over the problems with him.

Some indications that a market review is needed are:

- declining sales

- profit margins being squeezed

- fewer customers and over-reliance on a few large ones

- profitability not spread evenly over the product or service range - some items 'not earning their keep' - perhaps through being out of date, too expensive, badly designed, poorly constructed, etc.

Unfortunately some firms which close down do not seek financial advice until it is too late to halt the downward trend when earlier attention to the problems may have saved some of them. There are many reasons for this and one of them is that those running the business are unable to recognize the tell-tale signs and very few accountants take the trouble to explain to their clients what to look for. There are some tests and checks that can be done quite easily.

Cashflow

Cashflow is the lifeblood of the business and more businesses fail through lack of cash than for any other reason. Cash is generated through the conversion of work into debtors and then into payment and throught the deferral of the payment of supplies for as long a period as can be negotiated.

The objective must be to keep stock, work in progress and debts to a minimum and creditors to a maximum. Trends in important ratios as well as absolute values can help in assessing the business performance.

Debtor days

This is calculated by dividing your trade debtors by annual sales and multiplying by 365. It shows the number of days' credit being afforded to your customers and should be compared both with your normal trade terms and the previous month's figures. Normal procedures should involve the preparation of a monthly aged list of debtors showing the name of the customer, the value and which month it relates to.

The oldest and largest debtors can be seen at a glance for immediate consideration of what further recovery action is needed. The list may also show over-reliance on one or two large customers or the need to stop supplying a particularly bad payer until his arrears have been reduced to an acceptable level. Consideration should be given to making up bills to a date before the end of the month and making sure the accounts are sent out immediately followed by a statement 4 weeks later. Consider giving discounts for prompt payment.

If all else fails, and legal action for recovery is being contemplated, call at the County Court and ask for their leaflets, numbers 1 to 4.

Stockturn

The level of stock should be kept to a minimum and the number of days' stock can be calculated by dividing the stock by the annual purchases and multiplying by 365. A worsening trend on a month by month basis shows the need for action. It is important to make regularly a full inventory of all stock and dispose of old or surplus items for cash. A stock control procedure to avoid stock losses and to keep stock to a minimum should be implemented.

Profitability

Whilst cash is vital in the short-term, profitability is vital in the medium-term. The two key percentage figures are the gross profit percentage and the net profit percentage. Gross profit is calculated by deducting the cost of materials and direct labour from the sales figure whilst net profit is arrived at after deducting all overheads.

Possible reasons for changes in the gross profit percentage are:

- the pricing of jobs is not taking full account of increases in materials and wages and a review of pricing policy is needed

- too generous discount terms are being offered

- poor management, overmanning, waste and pilferage of materials

- too much down-time on plant which is in need of replacement.

If net profit is deteriorating after the deduction of an appropriate reward for your own efforts, including an amount for your own personal tax liability, you should review each item of overhead expenditure in detail asking amongst others the following questions:

- can savings be made in non-productive staff?

- is sub-contracting possible and would it be cheaper?

- have all possible energy-saving methods been fully explored?

- do the company's vehicles spend too much time in the yard; can they be shared and their number reduced?

- is the expenditure on advertising producing sales (review in association with 'marketing' above)?

Over-trading

Many inexperienced businessmen imagine that profitability equals money in the bank. In some cases, particularly where the receipts are wholly in cash, this may be the case, but additional business may mean higher stock inventories, extra wages, overheads, increased capital expenditure on premises and plant.

If the debtors show a marked increase as the turnover rises the proprietor may find to his surprise that each expansion of trade reduces rather than increases his cash resources. The business, which had enough funds for start-up, finds it does not have sufficient cash to run at the higher level of operation and the bank manager may be getting anxious about the increasing overdraft.

It is essential for those who run a business which operates on credit terms to be aware that profitability does not necessarily mean increased cash availability. Regular monthly management information on marketing and finance as described in this chapter will enable 'over-trading' to be recognized and remedial action to be taken early.

If the situation is appreciated only when the bank and other creditors are pressing for money, radical solutions may be necessary such as bringing in new finance, sale and leaseback of premises, a fundamental change in the terms of trade, or even selling out to a buyer with more resources. Professional help from the firm's accountant will be needed, whilst the TEC has counsellors experienced in advising on both the marketing and financial aspects of such situations.

Break-even point

The costs of a business may be divided into two types: variable and fixed.

Variable costs are those which increase or decrease as the volume of work goes up or down and include such items as materials used, direct labour, power, machine tools, etc.

Fixed costs are not related to turnover and are sometimes called 'fixed overheads'. They include rent, rates, insurance, heat and light, office salaries, plant depreciation, etc.; these costs are still incurred even though few or no sales are being made.

RUNNING THE BUSINESS

Many small businessmen run their enterprises from home using family labour as back-up; they sell mainly their own labour and buy materials and hire plant only as required.

By these means they reduce their fixed costs to a minimum and start making profits almost immediately. However, larger firms which have business premises, perhaps a small workshop, an office, vehicles, etc., need to know how much they have to sell to cover their costs and become profitable.

In the case of a new business it is necessary to estimate the figures but where annual accounts are available a break-even chart based on them can be readily prepared.

Suppose the real or estimated figures (expressed in £000s) are:

	%	£000
Sales	100	400
Variable costs	66	265
Gross profit	34	135
Fixed costs	13	50
Nett profit	21	85

Break-even point $=$ Fixed Costs $\div$ 1 - $\dfrac{\text{Variable Costs}}{\text{Sales}}$

$$= \ 50 \div \ 1 - \frac{265}{400}$$

$$= \ 50 \div \ (1 - 0.6625)$$

$$= \ 50 \div \ 0.3375$$

$$= \ £148 \ (\text{thousand})$$

In practice things are never quite as clear-cut as the figures above show, but nevertheless this is a very useful tool for assessing not only the break-even point but also the approximate amount of loss or profit arising at differing levels of turnover and also for considering pricing policy.

Chapter 3
Taxation

A new basis of charging tax on business profits took effect from 6 April 1994 for new businesses and from 6 April 1996 for businesses already established at 6 April 1994.

For new businesses there may still be an advantage in having an accounting date early in the tax year, that is shortly after 6 April. The advantage, however, is less than it would have been under the old rules and professional advice will be needed to choose the best accounting date for tax purposes.

For existing businesses at 6 April 1994 the way income tax is charged on business profits is to change radically. For the 1997/98 tax year onwards profits assessed for tax in a given tax year will be the profits of the accounting period ending in that year and not as now those of the preceding year.

Transitional rules will apply for the tax year 1996/97 whereby the profits of two accounting periods will be used to give an average. For example, a self-employed business with 30 April year end, the average period began on 1 May 1994.

Because of the averaging of profits charged to tax, a tax planning opportunity will arise subject to anti-avoidance provisions introduced by the Inland Revenue to penalize the artificial movement of profit into the averaging period.

In particular the following areas may offer tax planning opportunities:

1. the purchase as opposed to the leasing of capital equipment to be used in the business

2. the introduction of a partner into the business

3. where a partnership is considering, for sound business reasons, re-financing by way of personal loans rather than a partnership loan.

Professional advice will be needed to take the maximum advantage of these opportunities.

The timing of a cessation of business will still be important particularly until the new system is fully in place. Professional advice should be sought on the timing and tax cost of ceasing business.

Together with changes in the way taxable profits will be measured, the Inland Revenue are introducing a new system of charging and collecting personal tax. Again from the tax year 1996/97 the burden of assessing tax will shift from the Inland Revenue to the individual tax payer. The main features of this new system are as follows:

1.   the onus is on the taxpayer to provide and complete information

2.   the taxpayer will have a choice: he can calculate and pay his tax liability at the same time as making his return and this will need to be done by 31 January following the end of the tax year. Alternatively he can send in his tax return much earlier and the Inland Revenue will calculate the tax to be paid on 31 January

3.   the important aspect to the new system is that if the return is late, or the tax is paid late, there will be automatic penalties imposed on the taxpayer.

Spouses in business

If spouses work in a business, perhaps answering the phone, making appointments, writing business letters, making up bills and keeping the books, they should be properly remunerated for it. Being a payment to a family member the inspector of taxes will be understandably cautious in allowing it in full as a business expense. The payment should be:

1.   actually paid to them, preferably weekly or monthly and in addition to any housekeeping monies

2.   recorded in the business book

3.   reasonable in amount in line with their duties and the time spent on them.

If the wages paid to them exceed £56.99 p.w. Class 1 employer and employee NIC becomes due and if they exceed £3,445 p.a. (assuming they have no other income) PAYE tax will also be payable.

It should also be noted that once small businesses are well established and the spouse's earnings are approaching the above limits, consideration may be given to bringing them in as a partner. This has a number of effects:

1. there is no longer a need to relate the spouse's income (which is now a share of the profits) to the work they do (if any)

2. they will pay Class 2 and Class 4 NIC instead of the more costly Class 1 contributions and PAYE will no longer apply to their earnings

3. but remember - as partners any assets they own are vulnerable to proceedings by partnership creditors.

Premises

Many small businessmen cannot afford to rent or buy commercial premises and run their enterprises from home using part of it as an office where the books and vouchers, clients' records, trade manuals, etc., are kept and estimates and plans are drawn up. In these circumstances a portion of the outgoings on the property may be claimed as a business expense.

Car expenses are usually split on a fractional mileage basis between business journeys, which are allowable, and private ones, which are not, and a record of each should be kept. If the business does work only on one or two sites or for only one main contractor the inspector may argue that the true base of operations is the work site not the residence and seek to disallow the cost of travel between home and work. It is tax-wise therefore, and sound business practice, to have as many customers as possible and not work for just one client.

Appeals against assessments. To businessmen, income tax assessments are an anathema. They should resist the temptation to tear them up or put them behind the clock and forget about them. All assessments should be checked for accuracy immediately and, if excessive, the instructions on the notice about making an appeal should be followed and the appeal sent to the Tax District that issued the assessment. The appeal should also show how much of the tax charged should be postponed because the assessment is too high.

If this is not done within 30 days of the issue of the notice, the assessment becomes final and the inspector (and the general commissioners if they are asked to adjudicate) may well not accept a late appeal. In this event the taxpayer has no alternative but to pay the tax as charged on the assessment which may be estimated or contain additions to the profits which have not been agreed by him and his accountant. If the appeal is to be made by the accountant check with him before the 30 days are up to ensure that it has been submitted.

Keep copies of all correspondence with the inspector and collector. Letters can be mislaid or fail to be delivered and it is essential to have both proof of what was sent as well as a permanent record of all correspondence.

TAXATION

Some useful statistics on Income Tax 1994/95

The current personal allowance for a single person is £3,445. The personal allowances for people aged 65 to 74 and over 75 years are £4,200 and £4,370 respectively. The married couple's allowance is £1,720 and £2,665 for a couple between the ages 65 and 74 and £2,705 for a couple over 75 years.

Rates of tax

The value of the married couple's allowance has been reduced for 1994/95 as it is now only worth a maximum 20% instead of saving tax at a person's top rate. The rates of tax for 1994/95 are as follows:

Lower rate: 20% on taxable income up to £3,000
Basic rate: 25% on taxable income between £3,000 and £23,700
Higher rate: 40% on taxable income over £23,700

Example: Married man: wife has no income. On earnings of £20,000 he would pay:

		£
Income		20,000
Less personal allowance		3,445
Taxable income		16,555
	£3,000 at 20% =	600
	£13,555 at 25% =	3,389
		3,989
Less married couple's allowance	£1,720 at 20%	344
Tax payable		3,645

On earnings of £30,000 the tax he would pay:

		£
	£3,000 at 20% =	600
	£20,700 at 25% =	5,175
	£2,855 at 40% =	1,142
Less married couple's allowance	£1,720 at 20%	344
	Tax payable =	6,573

TAXATION

Taxation of husband and wife

A married woman is treated in much the same way as a single person with her own personal allowance and basic rate band. Husband and wife each make a separate return of their own income and the Inland Revenue deals with each one in complete privacy - letters about the husband's affairs will be addressed only to him and about the wife's only to her (unless the parties indicate differently). The allowances and relief are dealt with as follows.

Personal allowance

Husband and wife each has one of these - for 1994/5 it is £3,445 each.

Married couple's allowance

The amount is £1,720. This is initially due to the husband but if his income is too small to use it all he may transfer the surplus part to his wife but once he has passed some allowances in this way he cannot change his mind and ask for them back. Married couples may choose, however, how they wish the allowance to be allocated between them, subject to the right of the wife to claim half the allowance if she wishes. Here are two examples:

Husband and wife both in employment; both earn £7,000 p.a.

1993/4: husband decides to keep all the married couple's allowance. The relief would be:

	Husband £	Wife £
Personal allowance	3,445	3,445
Married couple's allowance	1,720	-
	5,165	3,445

1994/5: wife decides she wants her half-share of the allowance. The relief would be:

	Husband £	Wife £
Personal allowance	3,445	3,445
Married couple's allowance	860	860
	4,305	4,305

31

TAXATION

Basic rate band

Husband and wife each have £3,000 chargeable at 20% and up to £20,000 at 25%.

Mortgage interest relief

The ceiling for relief is unchanged for 1994/5 at £30,000. The value of relief, however, has been reduced in a similar way to the married couple's allowance in that there is a maximum rate of relief of 20%. If the loan is in the name of one spouse only, that one gets all the relief due. If it is in the joint names of husband and wife it is allowed equally between them. However, if both agree they can *jointly* ask for relief to be divided in any proportion that they wish.

Business losses

These are allowed only against the income of the person who incurs the loss. For example, a loss in the husband's business cannot be set against the wife's income from employment.

Joint income

In the case of joint ownership by husband and wife of assets which yield income - bank and building society accounts, shares, rented property, etc. - the Revenue will treat the income as arising equally to both and each will pay tax on one half of the income. If however the asset is owned in unequal shares or by one spouse only and the taxpayer can prove this, then the shares to be taxed can be adjusted accordingly if a joint declaration is made to the tax office setting out the facts.

General

Special rules apply in the year of marriage or separation and divorce and on the death of the husband or wife. Contact your local tax office for information.

Capital Gains Tax (individuals 1994/5)

Where an asset is disposed of, the first £5,800 of the gain is exempt from tax. In the case of husbands and wives, each has a £5,800 exemption so if the ownership of the assets is divided between them, it is possible to claim exemption on gains up to £11,600 jointly in the tax year. Any remaining gain is chargeable as though it were the top slice of the individual's income; therefore, according to his or her circumstances it might be charged at 20%, 25% or 40%. Here is an example:

TAXATION

Husband: He is in business and is liable at 40% on his profits. He has £7,000 of chargeable capital gains less the £5,800 exemption sum which leaves £1,200. This is taxed at 40% to produce a sum due to the Inland Revenue of £480.

Wife: She has no income but also has £7,000 of chargeable gains. Her exemption of £5,800 leaves a taxable sum of £1,200 and as she has no taxable income and the gain is less than the lower band it is charged at 20% making £240 to be paid in tax.

Retirement relief may be due on the disposal of certain business assets after the age of 55 (or before that date where retirement is due to proven ill health). The maximum relief against capital gains is £250,000 plus one half of the gains between £250,000 and £1,000,000. A businessman contemplating retirement, or sale of business when aged 55 or over should consult his accountant *before* taking any steps and *before* changing his working pattern (e.g. going part-time).

Business entertainment

No relief is due for expenditure on business entertainment or on gifts to customers, whether they are from this country or overseas. However, the cost of small trade gifts not exceeding £10 per person in value is still admissible provided that the gift advertises the business and does not consist of food or tobacco.

Construction industry

An uncertificated sub-contractor in the construction industry will suffer tax at the basic rate of 25%. Any overpayment on account of the 20% rate may be reclaimed at the end of the tax year.

Dates tax due

Income Tax 1994/5

Earned income (such as trading profits) 50% on 1 January 1995 and a further 50% on 1 July 1995. Unearned income (such as rents and interest) due 1 January 1995.

Capital Gains Tax

Tax on gains in year ended 5 April 1994 is due on 1 December 1994 (or within 30 days of the issue of the notice of assessment). Tax on gains arising in the current tax year are due on 1 December 1995.

Self-employed NIC rates (from 6 April 1994)

Class 2 Rate - £5.65 per week. If earnings are below £3,200 p.a. averaged over the year ask the DSS about 'small income exception'; details are in leaflet NI 27A.

Class 4 Rate - Business profits up to £6,490 p.a.: NIL. Profits between £6,490 and £22,360 p.a.: 7.3% of the profit. There is no charge on profits over £22,360 p.a. so the maximum amount of Class 4 contributions is £1,158.51. One half of the Class 4 contribution is deducted from the profits for income tax purposes. Class 4 contributions are collected by the Inland Revenue along with the income tax due.

Corporation Tax (years ended 31/3/94 and 31/3/95)

For the year ended 31/3/95 corporation Tax is charged at 25% for profits up to £300,000. The ceiling for the previous year ended 31/3/94 was £250,000. Where the accounting date of a company is not 31/3/94, profits have to be apportioned on the time basis to the respective tax years. Profits exceeding £300,000 will be effectively charged at 35% up to £1,500,000 when the rate reduces to 33%. Companies can carry back trading losses for up to 3 years.

Capital allowances (depreciation) rates

Plant and machinery	25%	
Business motor cars - cost up to £12,000	25%	
- cost over £12,000	£3,000	(maximum)
Industrial buildings	4%	
Commercial and industrial buildings in Enterprise Zones	100%	

VAT

The standard rate is 17.5%. Lower limit for registration (from 11 March 1992) turnover per annum - £45,000.

Bad debts and VAT

Relief is available for debts over 6 months.

Chapter 4
Estimating

Pity the poor estimator! If a job goes well there will be a queue of agents, foremen, tradesmen, surveyors and buyers to take the credit. If a job loses money, however, the one person who will be left isolated to take the blame is the unsung hero in the construction industry - the estimator!

His art is highly dependent upon making a series of intelligent guesses to fill in the gaps of information not covered by the specification drawings and/or the bills of quantities. The quality of these guesses (or subjective judgements as the jargon would have it!) will often make the difference between a job making a profit or a loss.

It is possible, of course, to have too much information! There are many examples of a contractor carrying out the first phase of a contract but losing the second phase in open tender despite the fact of having the site set-up already there. The reason for the loss is usually due to having too much local knowledge gained from working on Phase 1 and including the cost of the known local risks in the second tender. Ignorance frequently prevails!

Apart from determining the contract sum at the tender stage, a properly prepared estimate can also be used for the following purposes:

1. Calculation of bonus targets;

2. Preparation of material schedules;

3. Analysis of anticipated and actual costs;

4. Production of a programme;

5. Preparation of monthly valuation and the final account.

This book, however, is concerned only with the preparation of estimates at the tender stage. It should be recognized that estimating is a very imprecise art. This often surprises people outside the construction industry who imagine it to be '...merely a matter of counting bricks and pricing them' as someone once said to me. If only it was!

Even taking this comment at face value shows how inaccurate this view is. There are many different kinds of bricks all with a different value. The cost of sand and cement for the mortar varies from supplier to supplier as does the hire cost of a mixer. The labour costs are the most likely to show the greatest variation.

No two men work at the same pace and produce the same volume of work. So even an uncomplicated task such as building a wall can produce significantly different estimates. When it comes to more complicated work, the likelihood of different estimates producing wide variations in their tender increases proportionately.

The main divisions in a properly prepared estimate are:

1. Own work which can be sub-divided into
   (a) Labour
   (b) Materials
   (c) Plant;

2. Work to be sub-let to sub-contractors;

3. Site overheads;

4. Office overheads;

5. Profit.

When an enquiry is received a contractor should decide very quickly which parts of the work he would sub-let if he was awarded the contract. He should then send the relevant extracts from the enquiry documents (specification/bills of quantities/drawings) to two or three sub-contractors whilst he is preparing the remainder of the estimate.

Labour

An experienced contractor should know the net charge-out rate of the men he directly employs. The basic labour rate on which the information in this book is based is set out at the beginning of Chapter 5. Every contractor will probably have his own 'customized' version but the main items such as NI contribution, overtime payments, bonus, etc., must be included.

The particular circumstances of the job must also be considered. The work may have to be carried out outside normal trading hours or done in unpleasant working conditions - both these situations would produce higher labour costs.

ESTIMATING

Materials

Although it is usually possible to get a better price from another materials supplier, a contractor should weigh up whether it is worth spending a lot of his time investigating other sources of supply if the savings are marginal.

A supplier who delivers on time (including the occasional small item when necessary) and replaces defective materials without query is probably worth supporting even if his prices are slightly higher than some of his competitors. It is still worth checking other prices now and then, however, to make sure that they are not too far out of line; but reliability and service have a real value and should be recognized.

The most difficult aspect of pricing materials is making a realistic assessment for waste and theft. Materials which are bought in bulk and only part of the order is allocated to any particular job or operation usually attract the highest waste percentage which can be as high as 15 to 20%. Specialized 'one-off' items such as cylinders, heaters and the like are usually better looked after and the waste factor could be as low as 1%.

Theft is another problem and a prudent contractor should make some allowance in his estimates to cover for this frustrating part of the industry.

Plant

Most small contractors hire in plant as necessary so the estimator need only assess the time the plant will be required because the hire rate will be easily established. One point to watch out for is the question of minimum hire periods.

Even if a piece of equipment is only required for one hour, the full cost of a day's hire must be included in the rates if that is the minimum hire period that can be obtained.

Site overheads

The range of overheads to be provided on site will vary widely depending on the size and nature of the job. There are two main types of overheads - fixed and time-related. Examples of these will clearly reveal the difference between them.

Fixed overheads are those of a non-recurring nature such as the establishment of the site huts. The cost of this operation - say £500 - must be allowed for whether the job lasts 4 weeks or 104 weeks and is a fixed sum.

Time-related overheads however are totally linked to the time the facility or service is required. In the case of site hutting the estimator should allow a weekly cost for the anticipated contract period plus whatever time he thinks the huts will be needed for carrying out maintenance work.

The following should act as a checklist for both fixed and time-related overheads:

Site supervision
Site accommodation
Lighting
Water
Safety, health and welfare
Removal of rubbish
Cleaning
Drying out
Protection of work
Security
Small tools
Insurances
Travelling time and fares
Scaffolding
Temporary services
Temporary fencing and screens

Office overheads

All contractors should give careful thought to the cost of their office overheads and in particular should be aware of the relationship, expressed as a percentage, between these costs and turnover.

The following is an example of a small contractor's overheads. This list is not meant to be exhaustive.

			£
Rent		(52 x £50)	2,600
Printing, stationery, etc.			150
Telephone		(12 x £50)	600
Van	HP	(12 x £200)	2,400
	Insurance		400
	Tax		100
	Petrol	(52 x £30)	1,560
	Repairs	say	500
Carried forward			8,310

		£
Brought forward		8,310
Part-time clerk/typist (52 x 30 hours x £3.50)		5,460
Photocopier - lease	(52 x £25)	1,300
Word-processor - lease	(52 x £20)	1,040
Advertising	(52 x £20)	1,040
Membership of professional body		150
Sundries		1,000
		18,300

Let us assume that the firm's turnover (that is, the total amount billed for the previous year) is £100,000. A simple calculation of £18,300 x 100 ÷ £100,000 shows that the relationship of overheads to turnover is 18.3%. I have set out a table below showing how this percentage varies when the level of the overheads or the turnover changes.

Overheads £	Turnover £	%
18,300	100,000	18.30
18,300	80,000	22.90
18,300	120,000	15.25
24,000	100,000	24.00
30,000	80,000	37.50
12,000	120,000	10.00

It is important to know the current percentage so it can be added to each quotation (although not necessarily shown). Ask your accountant for the cost of the overheads for last year and assess the percentage to be added to the costs. Unless something dramatic happens mid-year this figure needs only be examined annually. An example which would affect the overheads total could be the employment of a non-working supervisor or the purchase of an extra vehicle. A cancellation of a project or an unexpected increase in work would also affect the turnover figure.

It can be seen, therefore, that although working on a notional percentage is acceptable you should keep your eye open for any circumstances which may make a significant alteration to it.

Profit

Profit is the difference between income and cost. Some newly established contractors find it puzzling that although they are certain that their work is profitable, it is not reflected in the bank balance! The simple explanation for this apparent paradox is that as soon as it is earned the profit is re-used to finance the next stage of work or the next job. It only appears as a tangible asset when either trading stops or after a series of profitable jobs are completed and paid for.

Expansion of the business (and it is very difficult for a successful contractor not to expand) will further delay the surfacing of the profit and even a well planned expansion programme will usually result in an increased overdraft. The profit is still in the business of course, although not in the form of cash in the bank but in debtors' accounts and work in progress.

Chapter 5
Rates for measured work

Generally

The rates contained in this chapter apply to contracts ranging from minor works
to those up to £50,000 and are based at May 1994 which means that material
prices are those current during the second quarter of 1994. The rates are exclusive
of VAT which may be chargeable depending on the nature of the work. These
rates are national average, and the following regional adjustments should be
made:

Scotland	+ 10%
Wales	- 4%
Northern Ireland	- 18%
England	
South West	- 8%
South East	+ 5%
Home Counties	+17%
Inner London	+20%
Outer London	+12%
East Anglia	+ 2%
Midlands	- 3%
North West	+ 2%
North East	- 3%

Item descriptions

These contain brief descriptions of items. Item descriptions for fittings are listed
so that the more commonly used fittings appear first. Each item includes for all
the work normally associated with that particular item even if it is not expressly
stated.

Labour hours

The time for fixing is expressed as a decimal fraction of an hour - 0.50 hours equals

30 minutes. The times are average and include for unloading, distributing and fixing in position.

Nett labour

The total labour cost of supplying and fixing is calculated by multiplying the total labour hours by the hourly rate.

The all-in rate is calculated in accordance with the rates of wages agreed and published by the Joint Industry Board for the Electrical Contracting Industry based upon the wages award that took effect from 1 March 1993. Although the rates for Technicians, Approved Electricians and Electricians are shown, all the rates in this book are based upon an average of an Electrician's and an Approved Electrician's rates, i.e. £9.57 per hour.

	Technician	Approved Electrician	Electrician
Net basic hourly rate	6.95	6.00	5.54
Travel time and travel allowances per hour (based on 15-20 miles each way)	1.76	1.61	1.53
Total basic hourly rate	8.71	7.61	7.07
Annual wages x 1862 hours	£16,218.02	£14,169.82	£13,164.34
Non-productive overtime (50 hours)	435.50	380.50	353.00
Public holidays 8 days x 7.5 hours	417.00	360.00	332.40
	£17,070.52	£14,910.32	£13,850.24
National Insurance	1,380.08	1,177.28	1,078.48
JIB Combined Benefits Scheme 52 weeks @ £24.11	1,253.72	1,253.72	1,253.72
Carried forward	£19,704.32	£17,341.32	£16,182.44

RATES FOR MEASURED WORK

Brought forward	£19,704.32	£17,341.32	£16,182.44
Severance pay @ 1.5%	295.57	260.12	242.74
Employers' liability and third party insurance say 2.5%	492.61	433.53	404.56
Total cost	£20,492.50	£18,034.97	£16,829.74
Cost per hour divided by 1,822 (1,862 less 40 hours' inclement weather)	£11.25	£9.90	£9.24

Note that nothing has been included for contributions to the Construction Industry Training Board (CITB) or Joint Training Limited (JTL).

Net material

The net material cost is the basic trade price available to a small electrical contractor less the discount indicated plus an allowance for waste. It may be possible to negotiate individual discounts greater than those indicated to reduce the figures shown.

Overheads/profit

Overhead charges are those costs which are not directly linked to any particular contract. They are usually expressed as a percentage of the nett cost of labour and materials which is calculated by dividing the total projected annual overhead cost by the projected net turnover. They are explained in more detail in Chapter 4.

Unit

This column gives the unit in which the item is normally measured. Conduits and cables are normally measured in linear metres designated 'm' and all other items are generally enumerated, designated 'nr'.

Total

The total column is the sum of the net labour, net material and overheads/profit columns.

RATES FOR MEASURED WORK

Builder's work

No allowances have been made for cutting holes, mortices, chases, etc., or moving floor coverings, furniture and taking up floorings.

PROJECT LABOUR FACTORS

Application

From the hours for each individual item of work, the total time in hours can be calculated. In addition to this time, adjustments for particular conditions can be made to ensure the correct time for the project is allowed. The increase in hours spent should be converted to a cost, adjusted for overheads and profit and added to the project cost to give a total figure.

Working without drawings

Occasionally, certain parts of the works have to be installed without adequately detailed drawings being available. The adjustment reflects the additional time required for sorting out problems, taking site measurements and confirming details with the clerk of works or site engineer, if one is in attendance on the project.

Working in a congested building/area

On almost every project, additional time will be spent due to having to work round other trades, equipment, furniture, etc. The adjustment reflects the additional time required to cater for such working conditions, especially where a high level of activity over a short period working alongside other trades is required.

Walking allowance

This allowance should be applied when the work is on a large site and the distance from the site huts/storage compounds to the place of work exceeds 80m. Additional allowances should be made for tall buildings, where operatives need to wait for the use of passenger lifts.

Working hours

The contractors may consider applying adjustment factors for the number of hours worked per week or for night or weekend work to offset payment of operatives' bonuses, or hourly rate increases. These adjustments must be at the discretion of the contractor, but factors have been given for guidance.

RATES FOR MEASURED WORK

Supervision

The time taken to carry out an installation can be affected by the amount of supervision provided by the client and the supervision needed due to the complexity and the size of the installation. The standard rates allow for a reasonable level of supervision by the client or his agents. The adjustments reflect the additions that should be applied for alternative conditions.

Winter working and exposed sites

The time taken for normal installation work is greatly affected by winter conditions, especially if the site is exposed. Besides the hire cost of good site accommodation, drying facilities, screens and lighting, adjustments to the basic rates should be applied. (In certain instances the main contractor may provide the necessary accommodation, etc.)

Interrupted working

Normally, there would be first and second 'fix' activities of installation work. An adjustment to the overall project hours should be made where it is necessary to have to carry out the installation work on several visits to the site.

High standard of workmanship

It must be assumed that the contractor would carry out his work to a high standard but there are times when a client is particularly demanding in the level of workmanship required and the necessary adjustments should be made.

Factories

When a factory is in operation, the installation can be affected by having to erect and move scaffolding or steps more often than necessary, comply with site safety precautions, or stop work for the normal operations of the factory.

Offices

Where the office is furnished the installation can be affected by having to work around and protect furniture and decorations.

Shops

If a shop is in operation, the installation can be affected by having to protect goods, move displays and wait while customers are being served.

RATES FOR MEASURED WORK

Dwellings

Where the dwelling is occupied, the installation is affected by having to move furniture and carpets, wait to gain access to the property and take measurements.

Project labour factors

		% Adjustment
Working without drawings		+ 25
Working in a congested building/area		+ 25
Walking allowance		
Large site with no transport	75m - 100m	+ 3
	100m - 250m	+ 5
	100m - 400m	+ 8
Working hours		
Hours per week	45 - 55	+ 8
	55 - 65	+ 15
	65 - 80	+ 25
Night work (for full week at a time)		+ 75
Weekend work		+ 50

Supervision
  Large site, poor quality drawings,
  non-repetitive work, increased
  complexity of installation - with resident engineer   + 3
                            - without resident engineer + 5

Extensive supervision required, compressed
installation programme, very complex
installation, large numbers of
operatives used   up to   + 10

Winter working and exposed sites	up to	+ 25
Interrupted working		+ 10
High standard of workmanship		+ 10
Factories in operation		+ 10
Offices furnished and occupied		+ 25
Shops in operation	open for business	+ 50
	goods/display cabinets exposed	+ 5
Dwellings furnished and occupied		+ 50
access not guaranteed		+ 10

THE YELLOW BOOK

<u>Electrical Materials Price Guide</u>

Produced as a convenient A5 pocket book, *The Yellow Book* lists over 2,700 current prices for the more than 2,600 electrical materials most commonly required, making it invaluable for the busy contractor.

Published 6 times a year to keep you constantly updated with the latest products and prices, it is an essential guide for estimating, tendering, checking invoices, specifying and buying.

The products listed include:-

Power cables, fire detection and alarm systems, cable support systems, conduit and trunking, ventilation systems, distribution equipment, switches and connectors, lighting fittings, lamps and tubes.

For more informed details please contact:-
Building Materials Market Research, Baden House,
7 St Peter's Place, Brighton BN1 6TB
Tel: (0273) 680041/2 Fax: (0273) 606588

Spon's Budget Estimating Handbook

2nd Edition

Tweeds Chartered Quantity Surveyors, UK

- provides back-of-the envelope calculations
- improved design therefore easier to use
- fine-tuned to suit market's needs

This is the only guide to concentrate entirely on approximate estimating. It contains a broad range of information to help the developer, quantity surveyor, engineer, architect and landscape architect to produce approximate costings to enable early decisions to be taken on the viability of proposed schemes.

This book provides a one-stop reference point in preparing estimates, so saving valuable time and increasing the quality of cost estimates.

Contents: Foreword. Preface to the first edition. Preface. Introduction. **Part One: Building work.** Square metre prices. Elemental costs. Composite rates. **Part Two: Civil engineering work.** Principal rates. Composite rates. Project costs. **Part Three: Mechanical and electrical work.** Square metre prices. Principal rates. **Part Four: Reclamation and landscaping.** Principal rates. Maintenance. **Part Five: Alterations and repairs.** Principal rates. **Part Six: General data.** Life cycle costing. The development process. Professional fees. Construction indices. Rebuilding costs. Index.

April 1994: 234x120: 272pp, 1 line illus
Hardback: 0-419-19250-6: £39.00

For further information and to order please contact: The Promotion Dept., E & F N Spon, 2-6 Boundary Row, London SE1 8HN
Tel: 071 865 0066 Fax: 071 522 9623

E & F N Spon

An imprint of Chapman & Hall

V
ELECTRICAL SUPPLY –
POWER AND LIGHTING SYSTEMS

	Unit	Labour hours	Net labour (£)	Net material (£)	O'heads /profit (£)	Total (£)
V21 GENERAL LIGHTING (Refer to Y60-61, 63, 71, 73-74 and 81)						
V40 EMERGENCY LIGHTING (Refer to Y60-61, 63, 71, 73-74 and 81)						
V41 STREET, AREA AND FLOOD LIGHTING (Refer to Y60-61, 63, 71, 73-74 and 81)						

V51 LOCAL ELECTRIC HEATING UNITS

General notes

1. A discount of 10% has been incorporated within the net material costs for convector heaters, oil filled radiators, portable fan heaters, warm air curtains, portable air conditioners and their accessories (Dimplex)

2. A discount of 5% has been incorporated within the net material costs for showers, water boilers, hand driers, storage water heaters, cistern fed heaters, immersion heaters and all accessories (Heatrae Sadia)

Panel convector heaters (Dimplex Heating Ltd); (fixed with and including brackets; taking down once for decoration; refixing)

	Unit	Labour hours	Net labour (£)	Net material (£)	O'heads /profit (£)	Total (£)
PLX075N/SPL 0.75kW	nr	1.50	14.36	50.03	9.66	74.05
PLX125N/SPL 1.25kW	nr	1.50	14.36	56.10	10.57	81.03
PLX150N/SPL 1.5kW	nr	1.50	14.36	61.01	11.31	86.68
PLX200N 2.0kW	nr	1.50	14.36	63.92	11.74	90.02
PLX300N 3.0kW	nr	1.75	16.75	76.24	13.95	106.94

LOCAL ELECTRIC HEATING UNITS

	Unit	Labour hours	Net labour (£)	Net material (£)	O'heads /profit (£)	Total (£)
With 24 hour timer; (fixed with and including brackets; taking down once for decoration; refixing)						
PLX075N/SPL/TI 0.75kW	nr	1.50	14.36	64.49	11.83	90.68
PLX125N/SPL/TI 1.25kW	nr	1.50	14.36	70.60	12.74	97.70
PLX150N/ SPL/TI 1.50kW	nr	1.50	14.36	76.10	13.57	104.03
PLX200N/TI 2.0kW	nr	1.50	14.36	79.94	14.14	108.44
PLX300N/TI 3.0kW	nr	1.75	16.75	90.93	16.15	123.83
Convector heaters; (fixed with and including brackets; taking down once for decoration; refixing)						
DX20S 2.0kW	nr	1.50	14.36	31.19	6.83	52.38
DX305 3.0kW	nr	1.75	16.75	40.52	8.59	65.86
With 24 hour programmable timer; (fixed with and including brackets; taking down once for decoration; refixing)						
DX205/TI 2.0kW	nr	1.50	14.36	49.26	9.54	73.16
DX25S/TI turbo 2.5kW	nr	1.50	14.36	93.18	16.13	123.67
DX30S/TI 3.0kW	nr	1.75	16.75	55.58	10.85	83.18
Skirting convector heater; (fixed with and including brackets; taking down once for decoration; refixing)						
Heater 0.5kW	nr	1.00	9.57	34.64	6.63	50.84
Connector kit	nr	0.50	4.79	4.82	1.44	11.05
Feet for heater	nr	0.50	4.79	4.50	1.39	10.68
Multi-purpose convector heaters						
Coldwatcher; (fixed with and including brackets; taking down once for decoration; refixing)						
MPH 500 0.5kW	nr	1.50	14.36	31.44	6.87	52.67
MPH 1000 1.0kW	nr	1.50	14.36	41.42	8.37	64.15

POWER AND LIGHTING SYSTEMS

	Unit	Labour hours	Net labour (£)	Net material (£)	O'heads /profit (£)	Total (£)
MKI Oil filled electric radiators (room temperature controlled); (fixed with and including brackets; taking down once for decoration; refixing)						
A38 0.5kW	nr	1.00	9.57	73.73	12.50	95.80
B48 0.75kW	nr	1.00	9.57	83.42	13.95	106.94
B310 0.75kW	nr	1.00	9.57	84.42	14.10	108.09
C412 1.0kW	nr	1.00	9.57	89.92	14.92	114.41
C220 1.0kW	nr	1.00	9.57	106.63	17.43	133.63
D416 1.5kW	nr	1.00	9.57	116.40	18.90	144.87
E420 2.0kW	nr	1.00	9.57	148.03	23.64	181.24
MKII Energy controlled models; (fixed with and including brackets taking down once for decoration; refixing)						
P075 0.75kW	nr	1.00	9.57	71.60	12.18	81.17
P100 1.0kW	nr	1.00	9.57	77.72	13.09	100.38
P150 1.5kW	nr	1.00	9.57	93.43	15.45	118.45
MKIII Energy controlled electric radiators; (fixed with and including brackets; taking down once for decoration; refixing)						
OFX075 0.75kW	nr	1.50	14.36	43.09	8.62	57.45
OFX100 1.0kW	nr	1.50	14.36	49.62	9.60	73.58
OFX150 1.5kW	nr	1.50	14.36	59.65	11.10	85.11
MKIII Energy controlled electric radiators with built-in 24 hour programmable timer; (fixed with and including brackets; taking down once for decoration; refixing)						
OFX075/T1 0.75kW	nr	1.50	14.36	56.18	10.58	81.12
OFX100/T1 1.00kW	nr	1.50	14.36	61.36	11.36	87.08
OFX150/T1 1.50kW	nr	1.50	14.36	73.49	13.18	101.03

LOCAL ELECTRIC HEATING UNITS

	Unit	Labour hours	Net labour (£)	Net material (£)	O'heads /profit (£)	Total (£)
Top plates for electric radiators and skirting convector heater; fixing to backgrounds requiring plugging for						
C412	nr	0.75	7.18	13.14	3.05	23.37
SCH5	nr	0.75	7.18	13.14	3.05	23.37
D416	nr	0.75	7.18	16.29	3.52	26.99
C220, E420	nr	0.75	7.18	19.65	4.02	30.85
Caster wheels						
RC 0291	nr	0.17	1.63	9.46	1.66	12.75
RC 9000	nr	0.17	1.63	7.31	1.34	10.28
Oil filled column radiator; (fixed with and including brackets; taking down once for decoration; refixing)						
BK10/TI 1.0kW	nr	1.00	9.57	78.15	13.16	100.88
7 day column radiator; (fixed with and including brackets; taking down once for decoration; refixing)						
BK24T 2.4kW	nr	1.00	9.57	162.29	25.78	197.64
Wall brackets for BK10/TI and BK24	nr	1.00	9.57	15.99	3.83	29.39
Real flame effect fires; free-standing; placing into position complete with fitted plug						
Hursley HUR20 2kW	nr	1.00	9.57	224.89	35.17	269.63
Lymington LYM28E 2.8kW	nr	1.00	9.57	273.25	42.42	325.24
Holbury HOL20 2.0kW	nr	1.00	9.57	268.77	41.75	320.09
Romsey ROM20 2.0kW	nr	1.00	9.57	290.96	45.08	345.61

POWER AND LIGHTING SYSTEMS

	Unit	Labour hours	Net labour (£)	Net material (£)	O'heads /profit (£)	Total (£)
Electronic fuel effect fires wallmounted; fixing to backgrounds including drilling, plugging and screwing						
Optima 2.0kW	nr	1.00	9.57	135.00	21.69	166.26
Rustic 2.4kW	nr	1.00	9.57	177.04	27.99	214.60
Fuel effect radiant fires free-standing; placing into position; complete with fitted plug						
Linwood FE20/C 2.0kW	nr	1.00	9.57	81.14	13.61	104.32
Minstead FE20/CA 2.0kW	nr	1.00	9.57	92.84	15.36	117.77
Electronic fuel effect radiant convector fires freestanding; placing into position; complete with fitted plug						
Theme teak 2.87kW	nr	1.00	9.57	166.40	26.40	202.37
Lyndhurst 430RCE 2.80kW	nr	1.00	9.57	166.40	26.40	202.37
Radiant focal point fires wall-mounted; fixing to backgrounds including drilling, plugging and screwing						
Langley FPL20E 2.0kW (with economizer control)	nr	1.50	14.36	138.80	22.97	153.16
Avon FPS20E 2.0kW (with economizer control)	nr	1.50	14.36	80.37	14.21	94.73
Wallmounted fan heaters; fixing to backgrounds including drilling, plugging and screwing						
WFC 3D 3.0kW	nr	1.75	16.75	104.40	18.17	139.32
WFE3/TI 3.0kW with 24 hour timer	nr	1.75	16.75	129.01	21.86	167.62

LOCAL ELECTRIC HEATING UNITS

	Unit	Labour hours	Net labour (£)	Net material (£)	O'heads /profit (£)	Total (£)
Tango wallmounted fan heater; fixing to backgrounds including drilling, plugging and screwing						
Tangoheat FX20 2.0kW	nr	1.50	14.36	29.91	6.64	50.91
Tangoshave FXS20 2.0kW	nr	1.50	14.36	55.29	10.45	80.10
Tangotime FXT20 2.0kW	nr	1.50	14.36	55.29	10.45	80.10
Tango Auto-time FX20/TI 2.0kW	nr	1.50	14.36	55.55	10.49	80.40
Quartz radiant heaters wallmounted fixing to backgrounds including drilling, plugging and screwing						
QX 1500 1.5kW	nr	2.00	19.14	134.60	23.06	176.80
QX 3000 3.0kW	nr	2.00	19.14	216.77	35.39	271.30
QX 4500 4.5kW	nr	2.00	19.14	299.93	47.86	366.93
Guards for quartz heaters; fixing to heater						
QX 1500 open mesh	nr	0.33	3.16	5.63	1.32	10.11
QX 1500 fine mesh	nr	0.33	3.16	10.98	2.12	16.26
QX 3000 open mesh	nr	0.33	3.16	9.47	1.89	14.52
QX 3000 fine mesh	nr	0.33	3.16	12.35	2.33	17.84
QX 4500 open mesh	nr	0.33	3.16	13.25	2.46	18.87
QX 4500 fine mesh	nr	0.33	3.16	13.87	2.55	19.58
Portable fan heaters freestanding; placing into postion; complete with fitted plug						
Tango 2 2.0kW	nr	0.50	4.79	35.11	5.98	45.88
Tango 3 3.0kW	nr	0.50	4.79	37.87	6.40	49.06
Bolero 3.0kW	nr	0.50	4.79	31.58	5.46	41.83
Cherry 2.0kW	nr	0.50	4.79	24.02	5.07	28.81
Figaro upright 2.4kW	nr	0.50	4.79	33.71	5.78	44.28
Warm air curtains						
AC3 3.0kW	nr	1.75	16.75	102.55	17.89	137.19
AC4S 4.5kW	nr	1.75	16.75	120.96	20.66	158.37
Remote control switch AC0001	nr	2.00	19.14	33.23	7.86	60.23

POWER AND LIGHTING SYSTEMS

	Unit	Labour hours	Net labour (£)	Net material (£)	O'heads /profit (£)	Total (£)
Base unit heater (beneath kitchen unit)						
BUH 24T 2.4kW	nr	2.50	23.93	107.57	19.72	151.22
Infra-red heaters						
IRM 750 0.75kW	nr	1.50	14.36	27.05	6.21	47.62
IRD 1000 1.0kW	nr	1.50	14.36	28.96	6.50	49.82
IR 1800 1.8kW	nr	1.50	14.36	32.81	7.08	54.25
IRX 750 0.75kW	nr	1.50	14.36	26.73	6.16	47.25
IRX 1000 1.00kW	nr	1.50	14.36	28.42	6.42	49.20
Radiant wall fires wallmounted; fixing to backgrounds including drilling, plugging and screwing						
Studio 2 2.0kW	nr	1.50	14.36	52.70	10.06	77.12
Studio 2 Super 2.0kW	nr	1.50	14.36	75.78	13.52	103.66
Studio 3 Super 3.0kW	nr	1.75	16.75	99.28	17.40	133.43
Studio Super Electronic 2.1kW	nr	1.50	14.36	120.32	20.20	154.88
Oil filled electric towel rails (chrome plated) wallmounted; fixing to backgrounds including drilling, plugging and screwing						
TRC 90/W 90W	nr	1.50	14.36	63.11	11.62	77.47
TRC 130/W 130W	nr	1.50	14.36	75.74	13.51	90.10
TRC 150/W 150W	nr	1.50	14.36	81.87	14.43	96.23
Electric towel rail dry element S50 45W	nr	1.50	14.36	36.10	7.57	58.03
Oil filled electric towel rails (stove enamelled) wallmounted; fixing to backgrounds including drilling, plugging and screwing						
TRS 120/W 120W	nr	1.50	14.36	47.79	9.32	71.47
TRS 175/W 175W	nr	1.50	14.36	53.96	10.25	78.57
TRS 200/W 200W	nr	1.50	14.36	57.78	10.82	82.96

LOCAL ELECTRIC HEATING UNITS

	Unit	Labour hours	Net labour (£)	Net material (£)	O'heads /profit (£)	Total (£)
Oil filled electric towel rails (brass plated with stored-on lacquer coating) wallmounted; fixing to backgrounds including drilling, plugging and screwing						
TRB130/W 130W	nr	1.50	14.36	90.70	15.76	120.82
Storage heaters (Dimplex range)						
Assembling components; fixing to background, plugging and screwing						
XL12N 1.7kW	nr	2.00	19.14	107.02	18.92	145.08
XL18N 2.6kW	nr	2.00	19.14	136.74	23.38	179.26
XL24N 3.4kW	nr	2.00	19.14	166.46	27.84	213.44
XLS12N 1.7kW	nr	2.00	19.14	113.63	19.92	152.69
XLS18N 2.6kW	nr	2.00	19.14	143.34	24.37	186.85
XL24N 3.4kW	nr	2.00	19.14	174.41	29.03	222.58
XLE12N 1.7kW	nr	2.00	19.14	137.82	23.54	180.50
XLE18N 2.6kW	nr	2.00	19.14	167.54	28.00	214.68
XLE24N 3.4kW	nr	2.00	19.14	197.28	32.46	248.88
XLT6 0.85kW	nr	1.00	9.57	70.31	11.98	91.86
Combined storage and connector heaters (Dimplex range)						
Assembling components fixing to backgrounds including drilling, plugging and screwing						
CXL12N 1.7kW	nr	2.00	19.14	142.70	24.28	186.12
CXL18N 2.6kW	nr	2.00	19.14	171.76	28.63	219.53
CXL24N 3.4kW	nr	2.00	19.14	206.12	33.79	259.05
CXLE12N 1.7kW	nr	2.00	19.14	173.50	28.90	221.54
CXLE18N 2.6kW	nr	2.00	19.14	202.56	33.26	254.96
CXLE24N 3.4kW	nr	2.00	19.14	236.92	38.41	294.47

POWER AND LIGHTING SYSTEMS

	Unit	Labour hours	Net labour (£)	Net material (£)	O'heads /profit (£)	Total (£)
Fan storage heaters (Dimplex range)						
Assembling components fixing to backgrounds including drilling, plugging and screwing						
XF28 4.0kW	nr	2.00	19.14	414.80	65.09	499.03
XF35 5.0kW	nr	2.00	19.14	484.48	75.54	579.16
XF42 6.0kW	nr	2.00	19.14	563.85	87.45	670.44
Shelves for storage heaters						
Fixing to backgrounds including drilling, plugging and screwing						
SHE 12	nr	1.00	9.57	14.32	3.58	27.47
SHE 18	nr	1.00	9.57	15.33	3.73	28.63
SHE 24	nr	1.00	9.57	17.04	3.99	30.60
Tamper proof control cover plates for storage heaters; fixing to heater						
XLN cover plate						
XL 9000 (metal)	nr	0.08	0.77	5.49	0.94	7.20
XL 9050 (plastic)	nr	0.08	0.77	5.49	0.94	7.20
CXLN cover plate						
CXL 9001 (metal)	nr	0.08	0.77	5.49	0.94	6.26
Portable air conditoners						
Freestanding; placing into position; complete with fitted plug						
DAC 3400 0.35kW with dehumidifier	nr	0.50	4.79	407.05	61.78	411.84
DAC 6300 0.69kW with heater and dehumidifier	nr	0.50	4.79	659.68	99.67	664.47

LOCAL ELECTRIC HEATING UNITS

	Unit	Labour hours	Net labour (£)	Net material (£)	O'heads /profit (£)	Total (£)
Electric heating (water heater and hand driers)						
Instantaneous showers 7.2kW models fixing to backgrounds including drilling, plugging and screwing						
Cameo plus	nr	1.70	16.27	66.80	12.46	95.53
Cameo standard	nr	1.70	16.27	70.20	12.97	99.44
Sapphire standard	nr	1.70	16.27	90.73	16.05	123.05
Carousel standard	nr	1.70	16.27	113.53	19.47	149.27
Solitaire standard	nr	1.70	16.27	143.36	23.94	183.57
Accolade electronic standard	nr	1.70	16.27	171.95	28.23	216.45
Instantaneous showers 8.5kW models fixing to backgrounds including drilling, plugging and screwing						
Sapphire standard	nr	1.70	16.27	98.33	17.19	131.79
Carousel standard with multi-function handset	nr	1.70	16.27	128.73	21.75	166.75
Solitaire standard	nr	1.70	16.27	153.43	25.45	195.15
Accolade standard	nr	1.70	16.27	185.25	30.23	231.75
Instantaneous showers 8.5kW models						
Carousel standard	nr	1.70	16.27	134.90	22.68	173.85
Accessories; fixing to showers						
Anti-vac valve	nr	1.00	9.57	18.91	4.27	32.75
Anti-vac valve and adapter	nr	1.25	11.96	22.52	5.17	39.65
Adapter for AVV	nr	0.25	2.39	3.77	0.92	7.08
Pumped instantaneous						
Fixing to backgrounds including drilling, plugging and screwing						
Sureflow 7.2kW	nr	1.70	16.27	197.60	32.08	245.95
Sureflow 8.2kW	nr	1.70	16.27	214.70	34.65	265.62

POWER AND LIGHTING SYSTEMS

	Unit	Labour hours	Net labour (£)	Net material (£)	O'heads /profit (£)	Total (£)
Power showers						
Fixing to backgrounds including drilling, plugging and screwing						
Supajet 100	nr	2.00	19.14	118.75	20.68	158.57
Supajet 100 recessed	nr	3.00	28.71	139.65	25.25	193.61
Supajet 200	nr	2.00	19.14	134.90	23.11	177.15
Water boilers; placing into position						
Slimline Urn 1.8kW freestanding; placing into position; complete with fitted plug	nr	0.50	4.79	56.05	9.13	69.97
Supreme 140 2.5kW wallmounted; fixing to backgrounds including drilling, plugging and screwing	nr	1.50	14.36	317.30	49.75	381.41
Supreme 155 2.5kW wallmounted; fixing to backgrounds including drilling, plugging and screwing	nr	1.50	14.36	379.05	59.01	452.42
Supreme 170 2.5kW wallmounted; fixing to backgrounds including drilling, plugging and screwing	nr	1.50	14.36	419.90	65.14	499.40
Hand wash wallmounted; fixing to backgrounds including drilling, plugging and screwing						
Concept complete with spray nozzle 3kW	nr	1.00	9.57	57.48	10.06	67.05
Handy 3 plastic spout 3kW	nr	1.00	9.57	56.53	9.91	76.01
Handy 3 chrome spout 3kW	nr	1.00	9.57	62.46	10.80	82.83
Handy 7 plastic spout 7kW	nr	1.00	9.57	60.80	10.56	80.93
Handy 7 chrome spout 7kW	nr	1.00	9.57	66.50	11.41	87.48
Accessory Handy 3 and 7 250mm chrome spout	nr	0.50	4.79	7.51	1.84	12.30
Hand driers wallmounted; fixing to backgrounds including drilling, plugging and screwing						
Handy Dri 14 1.4kW	nr	1.00	9.57	112.10	18.25	139.92
Handy Dri 19 1.9kW	nr	1.00	9.57	133.00	21.39	163.96

LOCAL ELECTRIC HEATING UNITS

Hand driers (cont'd)	Unit	Labour hours	Net labour (£)	Net material (£)	O'heads /profit (£)	Total (£)
Handy Dri 14E 1.4kW	nr	1.00	9.57	149.15	23.81	182.53
No Touch 2.3kW	nr	1.00	9.57	249.85	38.91	298.33
Hair Drier 1.0kW	nr	1.00	9.57	121.60	19.68	131.17

Storage water heaters

Point of use wallmounted; fixing
to backgrounds including drilling,
plugging and screwing

Express 84 7 litres 1kW	nr	1.60	15.31	89.54	15.73	120.58
Express 84 7 litres 3kW	nr	1.60	15.31	89.54	15.73	120.58
Express 84 15 litres 3kW	nr	1.60	15.31	160.55	26.38	202.24
UTC undersink 1.5 or 3kW	nr	1.60	15.31	218.50	35.07	268.88
UTC feet	nr	0.50	4.79	11.78	2.49	19.06

Oversink fixing to backgrounds
including drilling, plugging and
screwing

B3 (NP) 23 litre 3kW	nr	1.80	17.23	369.55	58.02	444.80
C3 (NP) 55 litre 3kW	nr	2.20	21.05	470.25	73.69	564.99
D3 (NP) 68 litre 3kW	nr	2.25	21.53	513.00	80.18	614.71
Telescope spout for Express 84, B, C, D units	nr	0.00	0.00	23.99	3.60	23.99

Streamline with spout and valve
wallmounted; fixing to backgrounds
including drilling, plugging and
screwing

Streamline 7 litre 1kW	nr	1.60	15.31	101.41	17.51	134.23
Streamline 7 litre 3kW	nr	1.60	15.31	101.41	17.51	134.23
Streamline 10 litre 1kW	nr	1.60	15.31	125.16	21.07	161.54
Streamline 10 litre 3kW	nr	1.60	15.31	125.16	21.07	161.54

Streamline heater only for use
with option packs fixing to
backgrounds including drilling,
plugging and screwing

Streamline 7 litre 1kW	nr	1.50	14.36	84.08	14.77	113.21
Streamline 7 litre 3kW	nr	1.50	14.36	84.08	14.77	113.21

POWER AND LIGHTING SYSTEMS

	Unit	Labour hours	Net labour (£)	Net material (£)	O'heads /profit (£)	Total (£)
Streamline 10 litre 1kW	nr	1.50	14.36	107.83	18.33	140.52
Streamline 10 litre 3kW	nr	1.50	14.36	107.83	18.33	140.52

Option packs for use with
streamline B, C, D and UTC

	Unit	Labour hours	Net labour (£)	Net material (£)	O'heads /profit (£)	Total (£)
Pack B standard spout valve	nr	0.50	4.79	17.34	3.32	25.45
Pack C telescope spout valve	nr	0.50	4.79	24.94	4.46	34.19
Pack D touch pad and spout	nr	1.00	9.57	68.16	11.66	89.39
Pack E touch pad and remote spout	nr	1.25	11.96	80.75	13.91	106.62
Pack G concealed services kit	nr	1.50	14.36	16.15	4.58	35.09
Pack H mixer battery	nr	1.50	14.36	58.43	10.92	83.71
Pack J 2 hole mixer tap	nr	1.00	9.57	91.63	15.18	116.38
Pack K monobloc mixer tap	nr	1.00	9.57	47.98	8.63	66.18
Pack M elbow pillar taps (pair)	nr	1.00	9.57	262.20	40.77	312.54
Pack P elbow basin taps (pair)	nr	1.00	9.57	256.98	39.98	306.53
Pack Q 2 hole elbow mixer tap	nr	1.00	9.57	291.65	45.18	346.40
Pack R pillar taps (pair)	nr	1.00	9.57	95.95	15.83	121.35
Pack S basin tap (hot)	nr	0.50	4.79	47.03	7.77	59.59
Pack T basin tap (cold)	nr	0.50	4.79	29.45	5.14	39.38
Pack Z undersink adaptor unit	nr	1.50	14.36	8.09	3.37	25.82

Cistern fed heaters

Fixing to backgrounds including
drilling, plugging and screwing

Maximum head 9 metres

	Unit	Labour hours	Net labour (£)	Net material (£)	O'heads /profit (£)	Total (£)
UDB 91 91 litres 3kW	nr	2.50	23.93	560.98	87.74	672.65
UDB 136 136 litres 3kW	nr	2.70	25.84	641.25	100.06	767.15
JS 227/3 227 litres 3kW	nr	3.20	30.62	1100.10	169.61	1300.33
JS 227/6 227 litres 6kW	nr	3.20	30.62	1249.25	191.98	1471.85
JS 227/9 227 litres 9kW	nr	3.20	30.62	1259.70	193.55	1483.87
KS 272/3 272 litres 3kW	nr	3.20	30.62	1165.65	179.44	1375.71
KS 272/6 272 litres 6kW	nr	3.20	30.62	1310.05	201.10	1541.77
KS 272/9 272 litres 9kW	nr	3.20	30.62	1319.55	202.53	1552.70
NS 454/6 454 litres 6kW	nr	3.50	33.50	1690.05	258.53	1982.08
NS 454/12 454 litres 12kW	nr	3.50	33.50	1938.00	295.72	2267.22
NS 454/18 454 litres 18kW	nr	3.50	33.50	2048.20	312.25	2393.95
OS 545/6 545 litres 6kW	nr	3.50	33.50	1865.80	284.89	2184.19
OS 545/12 545 litres 12kW	nr	3.50	33.50	2032.05	309.83	2375.38
OS 545/18 545 litres 18kW	nr	3.50	33.50	2220.15	338.05	2591.70

LOCAL ELECTRIC HEATING UNITS

	Unit	Labour hours	Net labour (£)	Net material (£)	O'heads /profit (£)	Total (£)
Maximum head 18 metres fixing to backgrounds including drilling, plugging and screwing						
B3 (P)23 23 litres 3kW	nr	1.80	17.23	359.10	56.45	376.33
C3 (P)55 55 litres 3kW	nr	2.20	21.05	454.10	71.27	475.15
D3 (P)68 68 litres 3kW	nr	2.25	21.53	488.30	76.47	509.83
UDB 91X 91 litres 3kW	nr	2.50	23.93	624.15	97.21	648.08
UDB 136X 136 litres 3kW	nr	2.70	25.84	715.35	111.18	741.19
JS 227/3X 227 litres 3kW	nr	3.20	30.62	1248.30	191.84	1278.92
JS 227/6X 227 litres 6kW	nr	3.20	30.62	1396.50	214.07	1427.12
JS 227/9X 227 litres 9kW	nr	3.20	30.62	1407.90	215.78	1438.52
KS 272/3X 227 litres 3kW	nr	3.20	30.62	1292.95	198.54	1323.57
KS 272/6X 272 litres 6kW	nr	3.20	30.62	1435.45	219.91	1466.07
KS 272/9X 272 litres 9kW	nr	3.20	30.62	1463.00	224.04	1493.62
NS 454/6X 454 litres 6kW	nr	3.50	33.50	1849.65	282.47	1883.15
NS 454/12X 454 litres 12kW	nr	3.50	33.50	2092.85	318.95	2126.35
NS 454/18X 454 litres 18kW	nr	3.50	33.50	2335.10	355.29	2368.60
OS 545/6X 545 litres 6kW	nr	3.50	33.50	2025.40	308.83	2058.90
OS 545/12X 545 litres 12kW	nr	3.50	33.50	2502.30	380.37	2535.80
OS 545/18X 545 litres 18kW	nr	3.50	33.50	2579.25	391.91	2612.75
Cistern type (self-venting) fixing to backgrounds including drilling, plugging and screwing						
FBM 25 25 litres 3kW	nr	2.50	23.93	404.70	64.29	492.92
FBM 50 50 litres 3kW	nr	2.50	23.93	501.60	78.83	604.36
FBM 75 75 litres 3kW	nr	2.50	23.93	596.60	93.08	713.61
FBM 125 125 litres 6kW	nr	2.50	23.93	725.80	112.46	862.19
BT 23 23 litres 3kW	nr	3.00	28.71	362.90	58.74	450.35
BT 34 34 litres 3kW	nr	2.50	23.93	419.90	66.57	510.40
BT 55 55 litres 3kW	nr	2.50	23.93	497.80	78.26	599.99
BT 68 68 litres	nr	2.50	23.93	518.70	81.39	624.02
UDB/BT91 91 litres	nr	3.30	31.58	562.40	89.10	683.08
UDB/BT 136 136 litres	nr	3.00	28.71	679.25	106.19	814.15
UDB/BT and UBD wall brackets	nr	0.50	4.79	53.20	8.70	66.69

POWER AND LIGHTING SYSTEMS

	Unit	Labour hours	Net labour (£)	Net material (£)	O'heads /profit (£)	Total (£)
Unvented units (high pressure) with safety valve and expansion vessel kit, placing into position						
Quickflow 90 direct 3kW	nr	4.50	43.06	438.90	72.29	554.25
Quickflow 90 indirect 3kW	nr	4.50	43.06	533.90	86.54	663.50
Quickflow 144 direct 6kW	nr	4.60	44.02	499.70	81.56	543.72
Quickflow 144 indirect 3kW	nr	4.60	44.02	597.55	96.24	737.81
Quickflow 210 direct 6kW	nr	4.80	45.94	672.60	107.78	826.32
Quickflow 210 indirect	nr	4.80	45.94	768.55	122.17	936.66
Multipoint 10 litre 3kW	nr	2.00	19.14	224.91	36.61	280.66
Multipoint 15 litre 3kW	nr	2.00	19.14	257.69	41.52	276.83
Accessories						
Pressure limiter and strainer	nr	2.00	19.14	41.56	9.11	69.81
Expansion vessel, valve and check valve	nr	3.00	28.71	75.29	15.60	119.60
Thermostatic blending valve	nr	1.00	9.57	63.60	10.98	73.17

Storage tanks

Cistern fed (require 1kW or 3kW immersion heaters measured elsewhere); placing into position

Maximum head 9 metres

	Unit	Labour hours	Net labour (£)	Net material (£)	O'heads /profit (£)	Total (£)
Cirrus 91 91 litres	nr	2.50	23.93	358.15	57.31	439.39
Cirrus 114 114 litres	nr	2.60	24.88	391.40	62.44	416.28
Cirrus 136 136 litres	nr	2.70	25.84	428.45	68.14	522.43
Cirrus 182 182 litres	nr	3.00	28.71	505.40	80.12	614.23
Cirrus 227 227 litres	nr	3.20	30.62	586.15	92.52	709.29

Maximum head 18 metres

	Unit	Labour hours	Net labour (£)	Net material (£)	O'heads /profit (£)	Total (£)
Cirrus 91X 91 litres	nr	2.50	23.93	398.05	63.30	485.28
Cirrus 114X 114 litres	nr	2.60	24.88	437.00	69.28	531.16
Cirrus 136X 136 litres	nr	2.70	25.84	481.65	76.12	583.61
Cirrus 182X 182 litres	nr	3.00	28.71	562.40	88.67	679.78
Cirrus 227X 227 litres	nr	3.20	30.62	653.60	102.63	786.85
Nimbus 114 114 litres	nr	2.60	24.88	506.35	79.68	610.91

LOCAL ELECTRIC HEATING UNITS

	Unit	Labour hours	Net labour (£)	Net material (£)	O'heads /profit (£)	Total (£)
Nimbus 136 136 litres	nr	2.70	25.84	548.15	86.10	660.09
Nimbus 182 182 litres	nr	3.00	28.71	626.05	98.21	752.97
Nimbus 227 227 litres	nr	3.20	30.62	722.00	112.89	865.51
Accessories						
Cirrus/Nimbus wall brackets	nr	0.50	4.79	59.85	9.70	74.34
Immersion heaters (fitting to tank and electrical connection) 2 1/4" BSP head						
Twin heat 23", 27", 30" long	nr	0.70	6.70	24.84	4.73	31.54
Gold dot 11", 14" long 1, 2 or 3kW	nr	0.70	6.70	14.06	3.11	20.76
Gold dot 18" long 3kW	nr	0.70	6.70	14.49	3.18	21.19
Gold dot 23", 27" long 3kW	nr	0.70	6.70	14.58	3.19	21.28
Gold dot 30" long 3kW	nr	0.70	6.70	14.77	3.22	21.47
Gold dot 36" long 3kW	nr	0.70	6.70	15.25	3.29	21.95
Superloy 2, 11" and 14" long, 1, 2 or 3kW	nr	0.70	6.70	19.48	3.93	26.18
Superloy 2 18" long 3kW	nr	0.70	6.70	19.67	3.96	26.37
Superloy 2 23" long 3kW	nr	0.70	6.70	19.76	3.97	26.46
Superloy 2 27" long 3kW	nr	0.70	6.70	19.76	3.97	26.46
Superloy 2 30" long 3kW	nr	0.70	6.70	20.05	4.01	26.75
Superloy 2 36" long 3kW	nr	0.70	6.70	20.57	4.09	27.27
Maxistore 14" long 3kW	nr	0.70	6.70	33.01	5.96	39.71
Maxistore 27", 30", 36" long 3kW	nr	0.70	6.70	61.66	10.25	68.36
1 1/4" BSP heads						
UX2 2kW	nr	0.70	6.70	15.39	3.31	25.40
UX3 3kW	nr	0.70	6.70	16.48	3.48	26.66
Side entry circulators						
8630A 8.5" long 3kW	nr	1.00	9.57	30.40	6.00	39.97
8630B 8.5" long 3kW	nr	1.00	9.57	27.93	5.62	37.50
117 10" long 3kW	nr	1.00	9.57	23.94	5.03	33.51
107 10" long 3kW	nr	1.00	9.57	21.47	4.66	31.04

POWER AND LIGHTING SYSTEMS

	Unit	Labour hours	Net labour (£)	Net material (£)	O'heads /profit (£)	Total (£)
Industrial immersion heaters						
2" BSP head (rod type)						
5R111 1kW	nr	1.00	9.57	62.23	10.77	82.57
5R215 2kW	nr	1.00	9.57	63.65	10.98	84.20
5R311 3kW	nr	1.00	9.57	86.45	14.40	110.42
5R615 6kW	nr	1.00	9.57	90.73	15.04	115.34
5R926 9kW	nr	1.00	9.57	107.35	17.54	134.46
5R1234 12kW	nr	1.00	9.57	119.70	19.39	129.27
2 1/4" BSP head (rod type)						
6R111 11" long 1kW	nr	1.00	9.57	62.70	10.84	72.27
6R215 15" long 2kW	nr	1.00	9.57	64.13	11.05	73.70
6R311 11" long 3kW	nr	1.00	9.57	87.88	14.62	97.45
6R330 30" long 3kW	nr	1.00	9.57	102.13	16.75	111.70
6R411 11" long 4kW	nr	1.00	9.57	82.65	13.83	92.22
6R415 15" long 4kW	nr	1.00	9.57	83.60	13.98	93.17
6R615 15" long 6kW	nr	1.00	9.57	91.20	15.12	100.77
6R630 30" long 6kW	nr	1.00	9.57	99.28	16.33	108.85
6R642 42" long 6kW	nr	1.00	9.57	108.78	17.75	118.35
6R724 24" long 7kW	nr	1.00	9.57	98.80	16.26	108.37
6R916 16" long 9kW	nr	1.00	9.57	98.80	16.26	108.37
6R923 23" long 9kW	nr	1.00	9.57	101.65	16.68	111.22
6R926 26" long 9kW	nr	1.00	9.57	102.60	16.83	112.17
6R936 36" long 9kW	nr	1.00	9.57	105.93	17.32	115.50
6R1223 23" long 12kW	nr	1.00	9.57	115.90	18.82	125.47
6R1234 34" long 12kW	nr	1.00	9.57	120.65	19.53	130.22
2" BSP head (core type)						
5C219 19" long 2kW	nr	1.00	9.57	95.48	15.76	105.05
5C327 27" long 3kW	nr	1.00	9.57	105.00	17.19	114.57
5C438 38" long 4kW	nr	1.00	9.57	120.65	19.53	130.22
5C650 50" long 6kW	nr	1.00	9.57	144.40	23.10	153.97
2 1/4" BSP head (core type)						
6C219 19" long 2kW	nr	1.00	9.57	95.95	15.83	105.52
6C327 27" long 3kW	nr	1.00	9.57	105.93	17.32	115.50
6C438 38" long 4.5kW	nr	1.00	9.57	121.13	19.60	130.70
6C650 50" long 6kW	nr	1.00	9.57	145.35	23.24	154.92

W
COMMUNICATIONS, SECURITY AND CONTROL SYSTEM

TELECOMMUNICATIONS

	Unit	Labour hours	Net labour (£)	Net material (£)	O'heads /profit (£)	Total (£)
W10 TELECOMMUNICATIONS (in addition refer to Y60 and Y74)						
General notes						
1. All labour times on accessories allow for making cable connections						
2. A discount of 24% has been incorporated within the nett material costs of the telephone outlets (MK equipment)						
(Outlets MK Ltd)						
Line jack telephone outlets, white plastic fixed to backgrounds with screws (boxes not included refer to Y74)						
flush single master	nr	0.33	3.16	4.67	1.17	9.00
flush single secondary	nr	0.33	3.16	3.26	0.96	7.38
flush twin master	nr	0.45	4.31	8.35	1.90	14.56
flush twin secondary	nr	0.45	4.31	5.74	1.51	11.56
surface single master	nr	0.33	3.16	4.96	1.22	9.34
surface single secondary	nr	0.33	3.16	3.61	1.02	7.79
Line jack telephone outlets, flush satin brass or matt chrome finish, fixed to backgrounds with screws						
single master	nr	0.33	3.16	7.91	1.66	12.73
single secondary	nr	0.33	3.16	7.17	1.55	11.88
twin master	nr	0.35	3.35	13.64	2.55	19.54
twin master	nr	0.35	3.35	9.51	1.93	14.79
Extension kit						
1 compact double adaptor, 1 compact line jack unit 15 metres of cable plus clips, for attaching cable to walls and skirting	nr	1.00	9.57	7.11	2.50	16.68

COMMUNICATIONS

	Unit	Labour hours	Net labour (£)	Net material (£)	O'heads /profit (£)	Total (£)
PTC plugs, fixed to line and handset cords						
4 way line plug	nr	0.20	1.91	0.46	0.36	2.73
4 way handset plug	nr	0.20	1.91	0.35	0.34	2.60
6 way line plug	nr	0.25	2.39	0.57	0.44	3.40
6 way handset plug	nr	0.25	2.39	0.57	0.44	3.40
PTC adaptors						
dual outlet 6 way	nr	0.25	2.39	8.00	1.56	11.95
unit with 3m cord DSL grommet plus tags	nr	0.35	3.35	13.46	2.52	19.33
PTC coupler connects two 6 way plugs	nr	0.10	0.96	2.19	0.47	3.62
compact line jack unit	nr	0.25	2.39	1.42	0.57	4.38
PTC cord sets, 3m lengths						
Line cords terminated with a plug at each end						
line cord 4 way	nr	0.10	0.96	3.90	0.73	5.59
line cord 6 way	nr	0.10	0.96	4.32	0.79	6.07
Line cords terminated with a plug at one end and spade terminals at the other						
line cord and tags, 4 way	nr	0.22	2.11	3.36	0.82	6.29
line cord and tags, 6 way	nr	0.25	2.39	4.06	0.97	7.42
Telephone cable						
1/0.5mm tinned copper covered with 0.15mm white PVC outer sheath clipped to backgrounds						
1 No of pairs	nr	0.17	1.63	0.09	0.26	1.98
2 No of pairs	nr	0.17	1.63	0.13	0.26	2.02
3 No of pairs	nr	0.17	1.63	0.20	0.27	2.10
4 No of pairs	nr	0.17	1.63	0.23	0.28	2.14
6 No of pairs	nr	0.25	2.39	0.33	0.41	3.13

TELECOMMUNICATIONS

	Unit	Labour hours	Net labour (£)	Net material (£)	O'heads /profit (£)	Total (£)
10 No of pairs	nr	0.33	3.16	0.61	0.57	4.34
12 No of pairs	nr	0.33	3.16	0.70	0.58	4.44
20 No of pairs	nr	0.50	4.79	1.11	0.89	6.79
Flexible handset cords already complete with a 4 way plug each end						
6ft length	nr	0.12	1.15	2.80	0.59	4.54
12ft length	nr	0.12	1.15	4.00	0.77	5.92
25ft length	nr	0.12	1.15	5.42	0.99	7.56
PTC/PCC insertion and cutting tool	nr	0.00	0.00	14.96	2.24	17.20
Philips telephone and answering machines, plugged into telephone outlet						
D9033 compact telephone	nr	0.20	1.91	18.68	3.09	23.68
D9039 deluxe desktop telephone	nr	0.20	1.91	32.30	5.13	39.34
TD9046 Slimline Mercury compatible telephone	nr	0.20	1.91	29.78	4.75	31.69
TD9150 telephone with memory function and on hook dialling	nr	0.20	1.91	29.74	4.75	31.65
TD9152 Multi-feature professional telephone, LED display on hook dialling	nr	0.20	1.91	38.25	6.02	40.16
TD9335 compact standalone answering machine	nr	0.32	3.06	37.49	6.08	40.55
Brother FAX machines plugged into telephone line						
FAX160, personal fax	nr	0.32	3.06	529.00	79.81	611.87
FAX450, desktop fax	nr	0.32	3.06	599.00	90.31	602.06
FAX550M, desktop fax with integral memory	nr	0.32	3.06	749.00	112.81	864.87
FAX1400M, desktop fax with integral memory and high speed modem	nr	0.32	3.06	1139.00	171.31	1313.37

STAFF PAGING AND LOCATION

	Unit	Labour hours	Net labour (£)	Net material (£)	O'heads /profit (£)	Total (£)

W11 STAFF PAGING AND LOCATION

General notes

1. All labour times include for any drilling and fixing equipment to backgrounds

2. Labour times include making final cable connections to equipment

Nurse call equipment

Lamp indicator/master control panels 24V DC operation complete with power supply unit including wiring connections

	Unit	Labour hours	Net labour (£)	Net material (£)	O'heads /profit (£)	Total (£)
5 way	nr	2.50	23.93	186.95	31.63	242.51
10 way	nr	3.00	28.71	220.94	37.45	287.10
15 way	nr	3.50	33.50	254.93	43.26	331.69
20 way	nr	4.00	38.28	328.57	55.03	421.88
25 way	nr	4.50	43.06	362.56	60.84	466.46
30 way	nr	5.00	47.85	447.54	74.31	569.70
40 way	nr	6.00	57.42	521.18	86.79	665.39
50 way	nr	7.00	66.99	770.44	125.61	963.04

Repeater panel for use with master control panel including wiring connections

	Unit	Labour hours	Net labour (£)	Net material (£)	O'heads /profit (£)	Total (£)
5 way	nr	2.50	23.93	107.64	19.74	151.31
10 way	nr	3.00	28.71	130.30	23.85	182.86
15 way	nr	3.50	33.50	152.95	27.97	214.42
20 way	nr	4.00	38.28	175.62	32.09	245.99
25 way	nr	4.50	43.06	198.28	36.20	277.54
30 way	nr	5.00	47.85	266.26	47.12	361.23
40 way	nr	6.00	57.42	300.25	53.65	411.32
50 way	nr	7.00	66.99	515.52	87.38	669.89
mute tone control on repeater	nr	0.75	7.18	22.66	4.48	34.32
reset control on repeater panel	nr	0.50	4.79	23.10	4.18	32.07
built-in charger and 24V battery	nr	1.00	9.57	150.15	23.96	159.72

COMMUNICATIONS

	Unit	Labour hours	Net labour (£)	Net material (£)	O'heads /profit (£)	Total (£)
Lettering in place of numbers to identify rooms on panel						
fascia (per way)	nr	0.00	0.00	12.71	1.91	14.62
installer's name and address	nr	0.00	0.00	38.12	5.72	38.12
Cable - for wiring from master panel to repeater panel laid in trunking or drawn into conduit						
4 core	m	0.03	0.29	0.25	0.08	0.54
10 core	m	0.03	0.29	1.27	0.23	1.56
20 core	m	0.04	0.38	2.08	0.37	2.46
40 core	m	0.06	0.57	3.19	0.56	3.76
Accessories						
pull cord call unit without reassurance lamp	nr	0.50	4.79	8.66	2.02	15.47
pull cord call unit with reassurance lamp	nr	0.60	5.74	17.73	3.52	26.99
push button call unit without reassurance lamp	nr	0.50	4.79	9.24	2.10	16.13
push button call unit with reassurance lamp	nr	0.60	5.74	18.60	3.65	27.99
combined push button call/reset unit with reassurance lamp	nr	0.75	7.18	29.52	5.50	42.20
Pear push call switch						
without lead	nr	0.50	4.79	5.54	1.55	11.88
with 2m lead and plug	nr	0.60	5.74	11.50	2.59	19.83
Pear push call unit complete with 2m lead, plug, holder and wallmounting sockets	nr	1.25	11.96	19.49	4.72	31.45
Pear push call unit with reassurance lamp and pear push holder complete with 2m lead plug and wallmounting socket	nr	1.33	12.73	33.38	6.92	46.11

	Unit	Labour hours	Net labour (£)	Net material (£)	O'heads /profit (£)	Total (£)
Combined pear push call/reset unit with reassurance lamp, plug and holder and complete with 2m lead, plug and wallmounting socket	nr	1.33	12.73	44.12	8.53	56.85
Point of call reset unit	nr	0.50	4.79	16.40	3.18	24.37
Point of call reset/lamp unit	nr	0.50	4.79	19.64	3.66	28.09
Overdoor lamp unit	nr	0.50	4.79	10.05	2.23	17.07
Floor radiator unit	nr	0.50	4.79	28.59	5.01	38.39
24V remote buzzer unit	nr	0.50	4.79	8.43	1.98	13.22
Pocket paging equipment for use with lamp indicator system						
Complete kit including VHF transmitter aerial, built-in cover, supply unit, and one pocket bleep pager	nr	8.00	76.56	560.18	95.51	732.25
Additional pocket 'bleep' pager	nr	0.25	2.39	190.58	28.95	221.92
Replacement battery for pager	nr	0.12	1.15	5.78	1.04	7.97
'Airphone' 2-way speech intercom equipment						
Master control console with handset including connection						
10 way	nr	6.00	57.42	717.72	116.27	891.41
20 way	nr	6.33	60.58	832.02	133.89	1026.49
30 way	nr	6.66	63.74	1051.51	167.29	1282.54
40 way	nr	7.00	66.99	1144.33	181.70	1393.02
50 way	nr	8.00	76.56	2616.08	403.90	3096.54
24V DC power supply unit	nr	1.00	9.57	92.35	15.29	117.21
Sub-station unit desk or wall-mounted	nr	2.00	19.14	48.51	10.15	67.65

COMMUNICATIONS

	Unit	Labour hours	Net labour (£)	Net material (£)	O'heads /profit (£)	Total (£)
Sub-station unit with privacy button	nr	2.00	19.14	64.10	12.49	95.73
Sub-station flush mounted	nr	3.00	28.71	67.86	14.49	111.06
Pear push call extension complete with lead and holder	nr	0.50	4.79	21.83	3.99	30.61
Pull cord unit for remote wiring to sub-stations	nr	0.50	4.79	12.42	2.58	19.79
Push button call sub-station unit	nr	0.50	4.79	51.69	8.47	56.48
Pull cord call sub-station unit	nr	0.50	4.79	51.69	8.47	64.95
Point of call reset unit	nr	0.50	4.79	29.40	5.13	39.32
Corridor lamp unit	nr	0.50	4.79	17.39	3.33	25.51
Remote tone-call unit	nr	0.50	4.79	16.63	3.21	24.63
Pocket paging equipment for use with 'Airphone' speech intercom systems						
Complete kit comprising VHF transmitter, aerial, built-in PSU and interface and one pocket 'bleep' pager	nr	8.00	76.56	674.14	112.60	863.30
Additional pager	nr	0.25	2.39	190.58	28.95	221.92
Replacement batteries for pager	nr	0.12	1.15	5.78	1.04	7.97
Repeat indicator panel for use with 'Airphone' 2-way intercom master consoles including power supply unit and tone unit						
5 way	nr	2.50	23.93	277.59	45.23	346.75
10 way	nr	3.00	28.71	358.05	58.01	444.77
15 way	nr	3.50	33.50	415.80	67.39	516.69
20 way	nr	4.00	38.28	537.08	86.30	661.66
25 way	nr	4.50	43.06	600.60	96.55	740.21

Repeat indicator panel (cont'd)	Unit	Labour hours	Net labour (£)	Net material (£)	O'heads /profit (£)	Total (£)
30 way	nr	5.00	47.85	750.75	119.79	918.39
40 way	nr	6.00	57.42	872.03	139.42	1068.87
50 way	nr	7.00	66.99	1420.65	223.15	1710.79
60 way	nr	8.00	76.56	1553.48	244.51	1874.55
2-way switch kit for operating two master consoles in parallel	nr	2.00	19.14	179.03	29.73	227.90

Home Office approved nurse/ warden call equipment

Central equipment

	Unit	Labour hours	Net labour (£)	Net material (£)	O'heads /profit (£)	Total (£)
Master console complete with charger and battery	nr	12.00	114.84	1308.62	213.52	1636.98
Transmitter/encoder assembly for use with console complete with aerial	nr	16.00	153.12	1625.86	266.85	1778.98
Mimic repeater panel display and tone only	nr	8.00	76.56	713.79	118.55	908.90
Printer for use with master console to record all call and reset incidents	nr	2.00	19.14	317.63	50.52	387.29

Portable receiver

	Unit	Labour hours	Net labour (£)	Net material (£)	O'heads /profit (£)	Total (£)
Pocket data pager with digital read-out	nr	0.25	2.39	294.53	44.54	341.46
5-bay charging pack complete with power supply unit	nr	0.50	4.79	242.55	37.10	284.44

Room units

Fixed room units

	Unit	Labour hours	Net labour (£)	Net material (£)	O'heads /profit (£)	Total (£)
call button and reset	nr	0.50	4.79	98.18	15.45	118.42
call and reset pull cord	nr	0.50	4.79	98.18	15.45	118.42
call pear-push and reset	nr	0.50	4.79	115.50	18.04	138.33
2 call pear-push and reset	nr	0.75	7.18	132.83	21.00	161.01
call button and emergency button and reset	nr	0.75	7.18	109.73	17.54	134.45

COMMUNICATIONS

	Unit	Labour hours	Net labour (£)	Net material (£)	O'heads /profit (£)	Total (£)
call pear-push and emergency button and reset	nr	0.75	7.18	121.28	19.27	147.73
2 call pear-pushes, emergency button and reset	nr	0.75	7.18	138.60	21.87	167.65
call pear-push, assistance button, emergency button and reset	nr	0.75	7.18	132.83	21.00	140.01
2 call pear-pushes assistance button, emergency button and and reset	nr	0.75	7.18	203.13	31.55	210.31
Portable room unit call button and rest	nr	0.25	2.39	121.28	18.55	142.22
Neck pendant call unit call push and reset	nr	0.25	2.39	101.06	15.52	118.97
Accessories 4dB loft antenna	nr	2.50	23.93	124.16	22.21	148.09
uniradio 67 50 ohm coaxial cable	nr	0.17	1.63	1.45	0.46	3.08
plug and adapter	nr	0.40	3.83	15.20	2.85	21.88
wallmounting bracket for portable room unit	nr	0.25	2.39	3.64	0.90	6.03
ceiling mounted slavepull cord call unit for use as a wired extension to all room units	nr	0.50	4.79	12.42	2.58	17.21
thermal printer paper 2 roll pack	nr	0.50	4.79	8.66	2.02	13.45
replacement 9V dry battery for all room units	nr	0.12	1.15	4.04	0.78	5.19
replacement mercury disposable battery for pocket data pager 3-cell pack	nr	0.12	1.15	3.18	0.65	4.98
overdoor lamp with (high brightness LED) for use with room call units	nr	0.33	3.16	17.33	3.07	23.56

STAFF PAGING AND LOCATION

	Unit	Labour hours	Net labour (£)	Net material (£)	O'heads /profit (£)	Total (£)
'Disabled toilet' alarm						
control panel including power supply, alarm indicators, alarm tone and reset button	nr	6.00	57.42	155.93	32.00	245.35
combined overdoor lamp/buzzer unit for use with disabled toilet alarm panel	nr	0.33	3.16	24.26	4.11	27.42

COMMUNICATIONS

	Unit	Labour hours	Net labour (£)	Net material (£)	O'heads /profit (£)	Total (£)

W20 TELEVISION
(in addition refer to Y60 and Y74)

General notes

1. All labour times on accessories connected to cables allow for termination and fixing

2. A 5% waste factor has been incorporated within the cable material costs

3. A discount of 24% has been incorporated within the nett material costs of the socket outlets and boxes (MK equipment)

Fitting, MK Ltd

TV/FM coaxial socket outlets flush white plastic fixed to backgrounds with screws (metal boxes not included refer to Y74)

	Unit	Labour hours	Net labour (£)	Net material (£)	O'heads /profit (£)	Total (£)
single outlet	nr	0.33	3.16	3.78	1.04	7.98
twin outlet	nr	0.42	4.02	5.43	1.42	10.87
isolated TV/FM single outlet	nr	0.35	3.35	6.60	1.49	11.44
twin outlet with FM/TV diplexer	nr	0.45	4.31	9.23	2.03	15.57

TV/FM coaxal socket outlets isolated. Flush satin brass or matt chrome finish, fixed to backgrounds with screws

	Unit	Labour hours	Net labour (£)	Net material (£)	O'heads /profit (£)	Total (£)
single outlet	nr	0.35	3.35	9.26	1.89	14.50

TELEVISION

	Unit	Labour hours	Net labour (£)	Net material (£)	O'heads /profit (£)	Total (£)
TV coaxial cable, laid in trunking or drawn in conduit						
Type-7/0.25mm plain copper stranded conductor, cellular polythene insulated and plain copper braided PVC sheath. General purpose TV aerial	nr	0.17	1.63	0.19	0.27	2.09
Type-1/1.12mm plain solid copper conductor, celluar polythene insulation, plain copper braiding covered by outer sheath of PVC. Low-loss TV aerial down lead	nr	0.17	1.63	0.54	0.33	2.17
Type-1/1.0mm plain solid conductor air spaced polythene insulation, plain copper braiding, covered with a brown PVC sheath	nr	0.17	1.63	0.40	0.30	2.03
TV coaxial cable connections						
Moulded plugs	nr	0.17	1.63	0.95	0.39	2.97
Sockets	nr	0.17	1.63	0.83	0.37	2.83
Aluminium plug	nr	0.17	1.63	0.55	0.33	2.51
Line connector	nr	0.30	2.87	1.52	0.66	5.05
Alternator	nr	0.30	2.87	3.29	0.92	7.08
Surface mounting outlet	nr	0.30	2.87	2.46	0.80	6.13
Aerial splitter/combiner						
low loss type	nr	0.20	1.91	8.13	1.51	11.55
resistive type	nr	0.20	1.91	5.91	1.17	8.99

COMMUNICATIONS

	Unit	Labour hours	Net labour (£)	Net material (£)	O'heads /profit (£)	Total (£)
TV aerials						
14 bay all channel UHF colour TV aerial including clamp	nr	1.00	9.57	42.38	7.79	59.74
8 bay all channel wideband colour TV aerial including clamp	nr	1.00	9.57	36.54	6.92	53.03
Accessories and fittings						
Automatic aerial rotator including control unit	nr	0.33	3.16	65.49	10.30	78.95
Rotator cable (30m) 3 core	nr	0.17	1.63	0.32	0.29	2.24
Rotator mast (1-5m)	nr	0.50	4.79	5.72	1.58	12.09
Guy wire galvanized steel (30m)	nr	0.17	1.63	0.14	0.27	1.77
Mast clamp	nr	0.17	1.63	2.52	0.62	4.77
Pack of 3 guy wire clamps	nr	0.17	1.63	1.09	0.41	3.13
U-bolt and clamp	nr	0.17	1.63	0.98	0.39	3.00
Wallmounting bracket	nr	0.33	3.16	16.45	2.94	19.61
1.2m wall mount	nr	0.33	3.16	5.16	1.25	8.32
Swayed 1.8m aerial mast	nr	0.50	4.79	5.49	1.54	11.82
Coaxial cable clip	nr	0.02	0.19	0.04	0.03	0.23

CLOCKS

	Unit	Labour hours	Net labour (£)	Net material (£)	O'heads /profit (£)	Total (£)
W23 CLOCKS (in addition refer to Y60, Y61 and Y74)						
General notes						
1. All labour times allow for any drilling and fixing clocks to backgrounds						
2. All labour times for mains operated clocks allow for making final cable connections						
3. All material costs in this section are nett trade. No discount has been applied						
Gent signal unity, 4 channel 7 days	nr	6.00	57.42	357.90	62.30	415.3
Regent analogue clocks - mains operated						
9" numeral	nr	0.50	4.79	44.35	7.37	56.5
9" cardinal	nr	0.50	4.79	36.96	6.26	48.0
9" ordinal	nr	0.50	4.79	44.35	7.37	56.5
9" 24 hour cardinal	nr	0.50	4.79	44.35	7.37	56.5
12" numeral	nr	0.50	4.79	48.70	8.02	61.5
12" cardinal	nr	0.50	4.79	40.35	6.77	51.9
12" ordinal	nr	0.50	4.79	48.70	8.02	61.5
12" 24 hour cardinal	nr	0.50	4.79	48.70	8.02	61.5
18" numeral	nr	0.50	4.79	119.10	18.58	142.4
18" cardinal	nr	0.50	4.79	119.10	18.58	142.4
18" ordinal	nr	0.50	4.79	119.10	18.58	142.4
18" 24 hour cardinal	nr	0.50	4.79	119.10	18.58	142.4
Regent analogue clocks - battery operated						
9" numeral	nr	0.50	4.79	40.10	6.73	51.62
9" cardinal	nr	0.50	4.79	33.39	5.73	43.9
9" ordinal	nr	0.50	4.79	40.10	6.73	51.62
9" 24 hour cardinal	nr	0.50	4.79	40.10	6.73	51.62
12" numeral	nr	0.50	4.79	44.40	7.38	56.57

COMMUNICATIONS

	Unit	Labour hours	Net labour (£)	Net material (£)	O'heads /profit (£)	Total (£)
12" cardinal	nr	0.50	4.79	36.96	6.26	48.01
12" ordinal	nr	0.50	4.79	44.40	7.38	56.57
12" 24 hour cardinal	nr	0.50	4.79	44.40	7.38	56.57
18" numeral	nr	0.50	4.79	119.10	18.58	142.47
18" cardinal	nr	0.50	4.79	119.10	18.58	142.47
18" ordinal	nr	0.50	4.79	119.10	18.58	142.47
18" 24 hour cardinal	nr	0.50	4.79	119.10	18.58	142.47
Gentime digital clocks						
model 991	nr	0.50	4.79	196.35	30.17	231.31
model 992	nr	0.50	4.79	446.10	67.63	518.52
model 993	nr	0.50	4.79	541.80	81.99	628.58
Gentime digital clocks mains operated						
model 991	nr	0.50	4.79	228.10	34.93	267.82
model 992	nr	0.50	4.79	471.30	71.41	547.50
model 993	nr	0.50	4.79	572.10	86.53	663.42
Gentime digital clocks mains operated						
model 991	nr	0.50	4.79	228.10	34.93	267.82
model 992	nr	0.50	4.79	471.30	71.41	547.50
model 993	nr	0.50	4.79	572.10	86.53	663.42
Mains digital clocks, agent range						
hours/mins, 24 hour red display	nr	0.50	4.79	230.60	35.31	270.70
hours/mins, 24 hours green display	nr	0.50	4.79	281.00	42.87	328.66
hours/mins/date/month either 12 or 24 hour green display	nr	0.50	4.79	570.80	86.34	661.93
hours/mins/date/month either 12 or 24 hour green display	nr	0.50	4.79	699.30	105.61	809.70
hours/mins/seconds either 12 or 24 hour red display	nr	0.50	4.79	331.40	50.43	386.62
hours/mins/seconds either 12 or 24 hour green display	nr	0.50	4.79	383.10	58.18	446.07

CLOCKS

	Unit	Labour hours	Net labour (£)	Net material (£)	O'heads /profit (£)	Total (£)
Clock connections 2 amp fuse, white plastic						
surface mounted fixed to backgrounds, drilling, plugging and screwing	nr	0.42	4.02	7.14	1.67	11.16
flush mounted fixed to backgrounds with screws (back box measured in Y74)	nr	0.28	2.68	7.84	1.58	10.52

COMMUNICATIONS

	Unit	Labour hours	Net labour (£)	Net material (£)	O'heads /profit (£)	Total (£)

W30 DATA TRANSMISSION
(in addition refer to Y60 and Y74)

General notes

1. All labour times on accessories
   allow for making final cable
   connections

2. A discount of 24% has been
   incorporated within the nett
   material cost for MK equipment

Cables laid in trunking or drawn
in conduit

Multi-conductor with foil screen
suitable for RS232 applications

	Unit	Labour hours	Net labour (£)	Net material (£)	O'heads /profit (£)	Total (£)
4 core	m	0.17	1.63	0.16	0.27	2.06
6 core	m	0.17	1.63	0.21	0.28	2.12
8 core	m	0.17	1.63	0.23	0.28	2.14
12 core	m	0.19	1.82	0.29	0.32	2.43
2 core 7/0.2mm screened misted pair	m	0.17	1.63	0.42	0.31	2.36
75 ohm coaxial cable	m	0.22	2.11	0.38	0.37	2.49

Multi-pair screened audio cable

	Unit	Labour hours	Net labour (£)	Net material (£)	O'heads /profit (£)	Total (£)
5 pair	m	0.22	2.11	1.22	0.50	3.83
10 pair	nr	0.24	2.30	2.46	0.71	5.47

Short universal cable type U
7/0.1mm

	Unit	Labour hours	Net labour (£)	Net material (£)	O'heads /profit (£)	Total (£)
	m	0.24	2.30	2.21	0.68	4.51

Beldon computer cable

	Unit	Labour hours	Net labour (£)	Net material (£)	O'heads /profit (£)	Total (£)
78 ohm axial	m	0.17	1.63	0.55	0.33	2.51
100 ohm twin axial	m	0.17	1.63	0.87	0.37	2.87
150 ohm twin axial	m	0.17	1.63	1.33	0.44	3.40
100 ohm two pair twin axial	m	0.20	1.91	1.70	0.54	4.15

DATA TRANSMISSION

	Unit	Labour hours	Net labour (£)	Net material (£)	O'heads /profit (£)	Total (£)
Plastic 'D' connectors, to cable tails, sockets						
9 way	nr	0.22	2.11	4.26	0.96	7.3.
15 way	nr	0.38	3.64	4.80	1.27	9.7
25 way	nr	0.80	7.66	5.88	2.03	15.5
37 way	nr	1.20	11.48	8.71	3.03	23.2
50 way	nr	1.40	13.40	13.23	3.99	30.6
Plugs						
9 way	nr	0.22	2.11	4.09	0.93	7.1
15 way	nr	0.38	3.64	4.62	1.24	9.5
25 way	nr	0.80	7.66	5.33	1.95	14.94
37 way	nr	1.20	11.48	8.00	2.92	22.4
50 way	nr	1.40	13.40	11.82	3.78	29.0
Metal shell 'D' connector backshells, to equipment						
9 way	nr	0.05	0.48	4.89	0.81	6.1
15 way	nr	0.05	0.48	5.17	0.85	6.5
25 way	nr	0.05	0.48	5.81	0.94	7.2
37 way	nr	0.05	0.48	7.33	1.17	8.98
Greenpar type N ethernet						
Connectors						
crimp plug	nr	0.17	1.63	4.55	0.93	7.11
solder jack	nr	0.17	1.63	4.64	0.94	7.21
resistor jack	nr	0.17	1.63	11.88	2.03	13.51
Adaptors						
straight	nr	0.17	1.63	12.02	2.05	15.70
N-BNC	nr	0.17	1.63	8.49	1.52	11.64
RJ11/RJ45 modular jacks						
IBM 3 pair	nr	0.20	1.91	3.28	0.78	5.97
IBM 3 pair with inside latch	nr	0.20	1.91	5.21	1.07	8.19
IBM 4 pair	nr	0.22	2.11	3.81	0.89	6.81

COMMUNICATIONS

	Unit	Labour hours	Net labour (£)	Net material (£)	O'heads /profit (£)	Total (£)
Shielded RJ 45 plugs						
round cable (24-28 AWG)	nr	0.17	1.63	0.90	0.38	2.91
flat cable (24-28 AWG)	nr	0.17	1.63	1.90	0.53	4.06
Fittings MK Ltd. Computer terminal sockets, flush white plastic, fixed to backgrounds with screws						
25 PIN 'D' type, single outlet	nr	0.28	2.68	6.16	1.33	10.17
25 PIN 'D' type, twin outlet	nr	0.42	4.02	11.00	2.25	17.27
BNC type, single outlet	nr	0.28	2.68	5.18	1.18	9.04
BNC type, twin outlet	nr	0.42	4.02	8.08	1.81	13.91
RJ 45 type, single outlet	nr	0.22	2.11	6.57	1.30	9.98
RJ 45 type, twin outlet	nr	0.36	3.45	10.72	2.13	16.30
601 W type, single outlet	nr	0.22	2.11	5.70	1.17	8.98
601 W type, twin outlet	nr	0.36	3.45	8.39	1.78	13.62
Computer terminal sockets satin brass and matt chrome finish, fixed to backgrounds with screws						
25 PIN 'D' type, single outlet	nr	0.28	2.68	10.19	1.93	14.80
25 PIN 'D' type, twin outlet	nr	0.42	4.02	15.97	3.00	19.99
BNC type, single outlet	nr	0.28	2.68	8.44	1.67	12.79
BNC type, twin outlet	nr	0.42	4.02	12.48	2.48	18.98
601 W type, single outlet	nr	0.22	2.11	8.50	1.59	12.20

	Unit	Labour hours	Net labour (£)	Net material (£)	O'heads /profit (£)	Total (£)

W40 ACCESS CONTROL
(in addition refer to Y30)

General notes

1. All labour times include all drilling, screwing and the like for fixing accessories to backgrounds

2. All labour times for wiring accessories allow for making final cable connections

3. All items allow for the installation to be carried out as whole systems. For individual items, adjust the time allowed

Audio door entry systems

Door panels in anodized aluminium complete with push buttons, name tag holders, speaker grille and flush backbox

	Unit	Labour hours	Net labour	Net material	O'heads/profit	Total
1 button	nr	0.80	7.66	26.08	5.06	38.80
2 buttons on 1 row	nr	1.00	9.57	29.32	5.83	44.72
3 buttons on 1 row	nr	1.00	9.57	35.48	6.76	51.81
4 buttons on 1 row	nr	1.00	9.57	40.12	7.45	57.14
6 buttons on 1 row	nr	1.20	11.48	49.38	9.13	60.86
8 buttons on 2 rows	nr	1.40	13.40	59.60	10.95	83.95
10 buttons on 2 rows	nr	1.60	15.31	68.65	12.59	83.96
12 buttons on 2 rows	nr	1.80	17.23	77.75	14.25	94.98
Universal audio amplifier (required with all door parts)	nr	1.00	9.57	38.50	7.21	55.28

COMMUNICATIONS

	Unit	Labour hours	Net labour (£)	Net material (£)	O'heads /profit (£)	Total (£)
Mahogany surface mounting frames or door panels						
1 and 2 button panels	nr	0.50	4.79	19.47	3.64	24.26
3 button panels	nr	0.50	4.79	20.72	3.83	29.34
4 button panels	nr	0.50	4.79	21.23	3.90	29.92
6 button panels	nr	0.50	4.79	22.45	4.09	31.33
8 and 10 button panels	nr	0.50	4.79	22.83	4.14	31.76
12 button panels	nr	0.50	4.79	23.40	4.23	32.42
28W power supply unit	nr	0.80	7.66	55.17	9.42	62.83
Apartment phone (ivory) complete with door release push button	nr	0.60	5.74	17.92	3.55	27.21
Electric door release for mortice locks	nr	2.00	19.14	15.45	5.19	39.78
Electric door release for rim locks	nr	2.00	19.14	15.94	5.26	40.34
8 core cable for use on door entry systems (100m reel)	nr	0.03	0.29	0.37	0.10	0.76
Video door entry systems						
Complete video kit for one-way system comprising door panel complete with 1 call button, speaker grille, amplifer, light, camera, lens and recessed back box. Power supply unit 4" wall video monitor and handset	nr	8.00	76.56	760.00	125.48	836.56

	Unit	Labour hours	Net labour (£)	Net material (£)	O'heads /profit (£)	Total (£)

W41 SECURITY, DETECTION AND ALARM (in addition refer to Y60)

General notes

1. All labour times include any drilling required and fixing equipment to backgrounds

2. Labour times include making final cable connections to equipment

Commercial intruder alarms (channel safety systems)

	Unit	Labour hours	Net labour (£)	Net material (£)	O'heads /profit (£)	Total (£)
Control panel to BS4737, 6 channel part select, built-in charger	nr	3.00	28.71	112.62	21.20	141.33
Control panel to BS4737 with remote key pad and tone unit, built-in charger	nr	2.50	23.93	225.23	37.37	249.16
Additional remote keypad	nr	1.50	14.36	63.53	11.68	89.57
Additional remote tone unit	nr	1.00	9.57	17.33	4.03	30.93
2V 6" internal alarm bell	nr	1.00	9.57	20.21	4.47	34.25
12V internal alarm siren	nr	1.00	9.57	20.21	4.47	34.25
Weatherproof self-activating bell and red housing	nr	1.00	9.57	77.96	13.13	100.66
Weatherproof self-activating siren and red housing	nr	1.00	9.57	77.96	13.13	100.66
Red weatherproof housing only	nr	1.00	9.57	28.88	5.77	44.22
12V flashing orange beacon	nr	1.00	9.57	21.37	4.64	30.94
Recessed magnetic door proximity switch	nr	1.20	11.48	4.91	2.46	18.85

COMMUNICATIONS

	Unit	Labour hours	Net labour (£)	Net material (£)	O'heads /profit (£)	Total (£)
Surface magnetic door proximity switch	nr	1.20	11.48	4.91	2.46	18.85
Heavy duty surface magnetic proximity switch	nr	1.20	11.48	6.93	2.76	21.17
Roller shutter magnetic proximity switch	nr	1.20	11.48	14.44	3.89	29.81
Stair pressure mat	nr	0.33	3.16	9.82	1.95	14.93
Floor pressure mat	nr	0.33	3.16	14.44	2.64	20.24
33mm roll of self-adhesive aluminium window foil	m	0.08	0.77	0.21	0.15	0.98
2-way make-off blocks	nr	0.08	0.77	1.10	0.28	1.87
Panic button, key reset	nr	0.50	4.79	9.82	2.19	16.80
Vibration window contact	nr	0.25	2.39	22.52	3.74	28.65
Instant trigger module	nr	0.50	4.79	11.03	2.37	18.19
Apollo 'D' passive infra-red detector 12m range	nr	0.80	7.66	54.86	9.38	62.52
By-pass key switch	nr	0.25	2.39	19.06	3.22	24.67
Door connector loop	nr	0.33	3.16	4.62	1.17	8.95
Terminal block - 8 way	nr	0.08	0.77	1.73	0.38	2.88
Terminal block - 6 way	nr	0.08	0.77	1.39	0.32	2.48
'999' police autodial unit including tape recording, power supply and 12V battery	nr	2.50	23.93	271.43	44.30	339.66
Programmable autodial unit including tape recording, power supply and 12V battery	nr	2.50	23.93	271.43	44.30	339.66

SECURITY, DETECTION AND ALARM

	Unit	Labour hours	Net labour (£)	Net material (£)	O'heads /profit (£)	Total (£)
Digital communicator including programming of E-prom, power supply and 12V battery	nr	3.00	28.71	288.75	47.62	365.08
Self-contained ultrasonic space alarm including built-in alarm siren	nr	1.00	9.57	167.69	26.59	177.26
Cable 1.5mm2 copper covered with a white PVC outer sheath. Clipped to backgrounds						
4-core security cable	m	0.17	1.63	0.23	0.28	2.14
6-core security cable	m	0.19	1.82	0.30	0.32	2.44
8-core security cable	m	0.21	2.01	0.37	0.36	2.74
12-core security cable	m	0.25	2.39	0.66	0.46	3.51
Vidor 12V dry battery	nr	0.06	0.57	11.27	1.78	13.62
Prechargeable 12V 1.2Ah sealed battery	nr	0.06	0.57	23.10	3.55	27.22
Spare key blank	nr	0.03	0.29	1.39	0.25	1.93
Industrial infra-red units						
Infra-red beam unit 15m range complete with mounting brackets	nr	2.00	19.14	153.04	25.83	198.01
Infra-red beam unit 110m range	nr	2.00	19.14	179.03	29.73	198.17
Weatherproof infra-red beam unit 40m range including heaters and mounting brackets	nr	2.00	19.14	306.08	48.78	374.00
12V lamp power supply unit including 5Ah rechargeable battery	nr	1.50	14.36	271.43	42.87	328.66

COMMUNICATIONS

	Unit	Labour hours	Net labour (£)	Net material (£)	O'heads /profit (£)	Total (£)
Domestic intruder alarms						
Superswitch - wireless intruder alarm system						
radio alarm system controllers						
external siren	nr	1.00	9.57	263.72	40.99	314.28
external siren	nr	1.00	9.57	169.51	26.86	205.94
external siren without light	nr	1.00	9.57	141.35	22.64	173.56
dummy external siren	nr	1.00	9.57	18.96	4.28	32.81
emergency transmitter for personal attack	nr	0.17	1.63	53.57	8.28	63.48
movement transmitter	nr	0.80	7.66	51.12	8.82	67.60
magnetic contacts	nr	1.20	11.48	1.70	1.98	15.16
small pressure mat	nr	0.33	3.16	2.75	0.89	6.80
large pressure mat	nr	0.33	3.16	2.80	0.89	6.85
Securitec - intruder alarm system						
Internal passive infra-red detectors						
pulse count	nr	0.80	7.66	19.47	4.07	31.20
mini pulse	nr	0.80	7.66	34.60	6.34	48.60
standard	nr	0.80	7.66	23.09	4.61	35.36
room alert (portable alarm, battery operated)	nr	0.80	7.66	43.31	7.65	58.62
chime alert (portable alarm, battery operated)	nr	0.80	7.66	14.96	3.39	26.01
External passive infra-red detectors						
double vision-twin heads	nr	0.80	7.66	61.66	10.40	79.72
red eye	nr	0.80	7.66	43.53	7.68	58.87
Outdoor PIR units						
to switch 500W	nr	0.80	7.66	27.71	5.31	40.68
400W	nr	0.80	7.66	32.17	5.97	45.80
master unit	nr	0.80	7.66	53.34	9.15	70.15
slave unit	nr	0.80	7.66	34.49	6.32	48.47
controller	nr	1.00	9.57	20.69	4.54	34.80

SECURITY, DETECTION AND ALARM

	Unit	Labour hours	Net labour (£)	Net material (£)	O'heads /profit (£)	Total (£)
External PIR halogen fittings						
red alert floodlight	nr	0.80	7.66	52.99	9.10	69.75
Powerflood - adjustable PLR	nr	0.80	7.66	35.75	6.51	49.92
Powerflood - nonadjustable PLR	nr	0.80	7.66	31.17	5.82	44.65
External PIR fittings						
Twinlite PIR maxload 500W (requires 2 x PAR38 120W lamps - not included)	nr	0.80	7.66	23.97	4.74	36.37
Securilite with additional switching capacity of up to 600W (tungsten)	nr	0.80	7.66	30.90	5.78	44.34
External photocell lamp fittings						
GL9	nr	0.80	7.66	26.17	5.07	38.90
Interceptorlite	nr	0.80	7.66	36.91	6.69	51.26
Globelite photocell	nr	0.80	7.66	21.93	4.44	34.03
Security spot light	nr	0.80	7.66	6.07	2.06	15.79
Photo-electric switch	nr	0.80	7.66	27.44	5.26	40.36
Intruder alarm panels						
2 zone panel	nr	2.00	19.14	32.06	7.68	58.88
3 zone panel	nr	2.00	19.14	36.37	8.33	63.84
5 zone panel	nr	2.50	23.93	75.37	14.89	99.30
2005 professional alarm kit (2 zone panel, bell box, siren, tamper switched, panic button, 3pr magnetic contacts, battery, PIR detectors, 50m of 6 core cable, instructions manual)	nr	8.00	76.56	103.90	27.07	180.46
Accessories						
pressure mat	nr	0.33	3.16	3.22	0.96	7.34
stair pressure mat	nr	0.33	3.16	2.18	0.80	6.14
personal attack alarm	nr	0.50	4.79	3.41	1.23	9.43
magnetic contacts (pair)	nr	1.20	11.48	0.72	1.83	14.03

COMMUNICATIONS

	Unit	Labour hours	Net labour (£)	Net material (£)	O'heads /profit (£)	Total (£)
bell box, complete with siren, self-activating bell module and microswitches	nr	1.00	9.57	34.64	6.63	44.21
siren	nr	0.80	7.66	8.03	2.35	18.04
dummy bell box	nr	0.80	7.66	12.02	2.95	22.63
microswitch	nr	0.80	7.66	7.51	2.28	17.45
re-chargeable 1.9 amp 12V battery	nr	0.17	1.63	13.53	2.27	17.43

FIRE DETECTION AND ALARM

	Unit	Labour hours	Net labour (£)	Net material (£)	O'heads /profit (£)	Total (£)
W50 FIRE DETECTION AND ALARM						
General notes						
1. For fire alarm cables refer to sections Y60 and Y61						
2. All labour times include for drilling and fixing equipment to backgrounds						
3. All labour times include making final cable connections to each piece of equipment						
4. All mateial costs in this section are nett trade - no discount has been applied						
Control panels						
1 zone self-contained panel c/w integral manual call point, electronic sounder and battery plus charger	nr	1.00	9.57	192.95	30.38	202.52
Manual fire alarm panels complete with batteries and charger						
1 zone	nr	1.00	9.57	112.00	18.24	139.81
2 zone	nr	1.50	14.36	155.00	25.40	194.76
4 zone	nr	2.50	23.93	275.50	44.91	344.34
8 zone	nr	4.50	43.07	397.50	66.09	506.66
8 zone	nr	4.50	43.07	397.50	66.09	506.66
8 zone repeat indicator panel	nr	1.00	9.57	119.70	19.39	148.66
5 zone panel with 5 alarm sectors	nr	4.00	38.28	992.25	154.58	1185.11
10 zone panel with 10 alarm sectors	nr	6.50	62.21	1256.85	197.86	1516.92

COMMUNICATIONS

	Unit	Labour hours	Net labour (£)	Net material (£)	O'heads /profit (£)	Total (£)
Addressable fire alarm panel c/w batteries and charger						
single loop panel with 16 zone capacity and LCD display	nr	1.50	14.36	498.00	76.85	589.21
flush shroud for semi-recessing	nr	0.30	2.87	43.00	6.88	45.87
repeat indicator panel with LCD display	nr	1.00	9.57	420.00	64.44	494.01
zone module kit PCB only for mounting in control panel or accessory box c/w end-of-line device	nr	0.50	4.79	49.00	8.07	61.86
accessory box for additional system PCB's relays and batteries	nr	0.40	3.83	99.00	15.42	102.83
Detectors						
common mounting base	nr	0.50	4.79	6.90	1.75	13.44
common mounting base and diode	nr	0.50	4.79	8.60	2.01	15.40
addressable detector base	nr	0.50	4.79	22.00	4.02	26.79
ionization smoke detector c/w mounting base	nr	0.53	5.07	45.15	7.53	50.22
optical smoke detector c/w mounting base	nr	0.53	5.07	42.90	7.20	55.17
fixed temperature heat detector (60 deg C) c/w mounting base	nr	0.53	5.07	23.50	4.29	28.57
rate of rise heat detector c/w mounting base	nr	0.53	5.07	23.50	4.29	32.86
fixed temperature heat detector (90 deg C) c/w mounting base	nr	0.53	5.07	23.50	4.29	28.57
LED remote detector indicator	nr	0.53	5.07	7.40	1.87	14.34
optical 3/4 wire smoke detector 24V c/w mounting base	nr	0.53	5.07	76.60	12.25	81.67
beam smoke detector	nr	0.50	4.79	254.30	38.86	297.95
duet detector 24V including optical smoke detector	nr	0.53	5.07	218.30	33.51	223.37
duet detector 24V c/w relay contacts and optical smoke detector	nr	0.53	5.07	270.60	41.35	275.67
duet detector 240V AC c/w relay contacts and optical smoke detector	nr	0.53	5.07	283.20	43.24	288.27

FIRE DETECTION AND ALARM

	Unit	Labour hours	Net labour (£)	Net material (£)	O'heads /profit (£)	Total (£)
Break glass manual call points						
flush mounted call point	nr	0.50	4.79	8.10	1.93	12.89
surface mounted call point	nr	0.50	4.79	8.60	2.01	15.40
addressable manual call point	nr	0.50	4.79	39.00	6.57	50.36
addressable key switch call point	nr	0.50	4.79	51.25	8.41	64.45
addressable manual call point weather resistant	nr	0.50	4.79	42.30	7.06	54.15
Alarm sounders						
12V DC 150mm electronic bell	nr	0.80	7.66	18.25	3.89	25.91
24V DC 150mm electronic bell	nr	0.80	7.66	14.50	3.32	22.16
24V DC 150mm electronic bell, weather resistant	nr	0.80	7.66	29.00	5.50	36.66
240V AC 150mm electronic bell	nr	0.80	7.66	24.19	4.78	31.85
240V AC 150mm electronic bell, weather resistant	nr	0.80	7.66	35.10	6.41	42.76
24V DC siren	nr	0.80	7.66	69.44	11.56	77.10
240V AC siren	nr	0.80	7.66	61.44	10.36	69.10
flashing xenon beacon 24V DC	nr	0.80	7.66	32.00	5.95	39.66
flashing xenon beacon 240V AC	nr	0.80	7.66	41.40	7.36	49.06
Battery chargers						
charger/audicle less cells 12V, 2 amp	nr	1.50	14.36	157.50	25.78	171.86
charger/cubide less cells 24V, 1.5 amp	nr	1.50	14.36	157.50	25.78	171.86
charger/cubide less cells 24V, 3 amp	nr	1.50	14.36	340.20	53.18	354.56
charger/cubide less cells 24V, 6 amp	nr	1.50	14.36	489.50	75.58	503.86
Batteries						
Fully sealed lead acid 24v, rechargeable						
6Ah (2 x 12V)	nr	0.17	1.63	63.40	9.75	65.03
12Ah (2 x 12V)	nr	0.17	1.63	126.90	19.28	128.53
24Ah (2 x 12V)	nr	0.17	1.63	191.90	29.03	193.53

COMMUNICATIONS

	Unit	Labour hours	Net labour (£)	Net material (£)	O'heads /profit (£)	Total (£)

W51 EARTHING AND BONDING

General notes

1. All prices of copper are based on the London Metal Exchange. Base rate of 1501-1600 pounds steriling

2. A discount of 10% has been included within the net material costs

3. 10% waste factor has been added to the material prices of conductor tapes and bars

4. Earth bars and disconnecting links mounted on insulators, fixed to backgrounds requiring drilling plugging and screwing

	Unit	Labour hours	Net labour	Net material	O'heads/profit	Total
Disconnecting link	nr	0.30	2.87	21.67	3.68	28.22
Six way earth bar with single disconnecting link	nr	0.75	7.18	80.01	13.08	100.27
Six way earth bar with twin disconnecting link	nr	0.75	7.18	89.36	14.48	111.02
Six way earth bar	nr	0.75	7.18	71.22	11.76	90.16
Swan-neck link	nr	0.10	0.96	3.26	0.63	4.85
Insulators fixed to backgrounds with nuts						
with 2 studs and 3 nuts	nr	0.20	1.91	3.32	0.78	6.01
insulator only	nr	0.20	1.91	2.83	0.71	5.45

EARTHING AND BONDING

	Unit	Labour hours	Net labour (£)	Net material (£)	O'heads /profit (£)	Total (£)
Air terminals, fixed to backgrounds						
Copper						
taper pointed 15 x 500mm	nr	0.50	4.79	7.83	1.89	14.51
taper pointed 15 x 1000mm	nr	0.50	4.79	14.32	2.87	21.98
taper pointed 15 x 2000mm	nr	0.50	4.79	26.17	4.64	35.60
air rod 10 x 500mm	nr	0.50	4.79	5.54	1.55	10.33
air rod 10 x 1000mm	nr	0.50	4.79	8.42	1.98	13.21
Aluminium						
taper pointed 15 x 500mm	nr	0.50	4.79	6.09	1.63	12.51
air rod 10 x 500mm	nr	0.50	4.79	4.28	1.36	9.07
Air terminal base						
copper rod diameter 15mm	nr	0.40	3.83	7.97	1.77	13.57
aluminium rod diameter 15mm	nr	0.40	3.83	5.68	1.43	10.94
copper rod diameter 10mm vertical	nr	0.40	3.83	7.30	1.67	12.80
aluminium rod diameter 10mm vertical	nr	0.40	3.83	5.81	1.45	11.09
copper rod diameter 10mm horizontal	nr	0.40	3.83	7.30	1.45	11.13
aluminium rod diameter 10mm horizontal	nr	0.40	3.83	5.81	1.45	11.09

Conductors fixed to backgrounds

Bare copper tape

Width mm	Thickness mm	Unit	Labour hours	Net labour (£)	Net material (£)	O'heads /profit (£)	Total (£)
12.5	1.5	m	0.27	2.58	0.91	0.52	4.01
12.5	3	m	0.27	2.58	1.83	0.66	5.07
20	1.5	m	0.27	2.58	1.45	0.60	4.63
20	3	m	0.27	2.58	2.42	0.75	5.75
25	1.5	m	0.27	2.58	1.86	0.67	5.11
25	3	m	0.27	2.58	3.03	0.84	6.45
25	4	m	0.27	2.58	4.13	1.01	7.72
25	6	m	0.27	2.58	6.03	1.29	9.90

COMMUNICATIONS

		Unit	Labour hours	Net labour (£)	Net material (£)	O'heads /profit (£)	Total (£)
31	3	m	0.30	2.87	3.83	1.00	7.70
31	6	m	0.30	2.87	7.68	1.58	12.13
38	3	m	0.33	3.16	4.71	1.18	9.05
38	5	m	0.33	3.16	7.84	1.65	12.65
38	6	m	0.33	3.16	9.45	1.89	14.50
50	3	m	0.35	3.35	6.20	1.43	10.98
50	4	m	0.35	3.35	8.25	1.74	13.34
50	6	m	0.35	3.35	12.05	2.31	17.71

PVC covered copper tape
Width   Thickness
mm   mm

12.5	1.5	Black	m	0.27	2.58	1.33	0.59	4.50
25	3	Black	m	0.27	2.58	4.39	1.05	8.02
25	6	Green	m	0.27	2.58	7.74	1.55	11.87
50	6	Green	m	0.35	3.35	16.24	2.94	22.53

Lead covered copper tape

25 x 3mm	m	0.27	2.58	17.14	2.96	22.68

LSF covered copper tape

25 x 3mm	m	0.27	2.58	7.36	1.49	11.43

Tinned copper tape

12.5 x 1.5mm	m	0.27	2.58	2.42	0.75	5.75
25 x 3mm	m	0.27	2.58	5.25	1.17	9.00
25 x 6mm	m	0.27	2.58	8.49	1.66	12.73

Hard drawn copper bar

25 x 3mm	m	0.30	2.87	6.93	1.47	11.27
25 x 6mm	m	0.30	2.87	13.57	2.47	18.91
38 x 6mm	m	0.30	2.87	21.56	3.66	28.09
50 x 6mm	m	0.35	3.35	27.22	4.59	35.16
50 x 10mm	m	0.35	3.35	46.35	7.46	57.16
75 x 6mm	m	0.35	3.35	41.23	6.69	51.27

Tinned bar

50 x 6mm	m	0.35	3.35	35.33	5.80	44.48

EARTHING AND BONDING

	Unit	Labour hours	Net labour (£)	Net material (£)	O'heads /profit (£)	Total (£)
Flexible copper braid						
15 x 2.4mm	m	0.27	2.58	3.44	0.90	6.92
25 x 3.5mm	m	0.27	2.58	7.15	1.46	11.19
Bare aluminium tape						
Width Thickness						
mm mm						
12.5 1.5	m	0.27	2.58	0.41	0.45	3.44
20 3	m	0.27	2.58	0.94	0.53	4.05
25 3	m	0.27	2.58	1.18	0.56	4.32
25 6	m	0.27	2.58	2.38	0.74	5.70
50 6	m	0.35	3.35	4.73	1.21	9.29
PVC covered aluminium tape						
Width Thickness						
mm mm						
12.5 1.5 Black	m	0.27	2.58	0.91	0.52	4.01
20 3 Black	m	0.27	2.58	2.28	0.73	5.59
25 3 Black	m	0.27	2.58	2.29	0.73	5.60
Bare solid circular conductor						
copper 8mm	m	0.20	1.91	2.00	0.59	4.50
aluminium 8mm	m	0.20	1.91	0.74	0.40	3.05
PVC covered solid						
Circular conductor 8mm diameter						
copper black	m	0.20	1.91	2.82	0.71	5.44
aluminium brown	m	0.20	1.91	1.51	0.51	3.42

COMMUNICATIONS

	Unit	Labour hours	Net labour (£)	Net material (£)	O'heads /profit (£)	Total (£)

Conductor fixings

Non-metallic DC clip (for use with bare tape)

Conductor size Colour
mm

20 x 3 Brown	nr	0.20	1.91	0.32	0.33	2.56
25 x 3 Brown	nr	0.20	1.91	0.32	0.33	2.56
38 x 6 Brown	nr	0.27	2.58	0.32	0.43	3.33
50 x 6 Brown	nr	0.27	2.58	0.83	0.51	3.92

For use with PVC covered tape

25 x 3mm	nr	0.20	1.91	0.35	0.34	2.60

Adhesive DC clip

for use with bare tape	nr	0.18	1.72	1.52	0.49	3.73
for use with PVC covered tape	nr	0.18	1.72	1.52	0.49	3.73

DC tape clip

Size mm   Conductor material

20 x 3 Bare copper	nr	0.20	1.91	0.95	0.43	3.29
25 x 3 Bare copper	nr	0.20	1.91	0.97	0.43	3.31
25 x 4 Bare copper	nr	0.20	1.91	1.14	0.46	3.51
25 x 6 Bare copper	nr	0.20	1.91	1.53	0.52	3.96
31 x 3 Bare copper	nr	0.20	1.91	2.13	0.61	4.65
31 x 6 Bare copper	nr	0.20	1.91	2.33	0.64	4.88
38 x 3 Bare copper	nr	0.20	1.91	2.45	0.65	5.01
38 x 5 Bare copper	nr	0.27	2.58	2.49	0.76	5.83
38 x 6 Bare copper	nr	0.27	2.58	2.54	0.77	5.89
50 x 3 Bare copper	nr	0.27	2.58	2.49	0.76	5.83
50 x 4 Bare copper	nr	0.27	2.58	2.54	0.77	5.89
50 x 6 Bare copper	nr	0.27	2.58	1.94	0.68	5.20
25 x 3 PVC covered copper	nr	0.20	1.91	2.47	0.66	5.04
25 x 6 PVC covered copper	nr	0.20	1.91	2.56	0.67	5.14
50 x 6 PVC covered copper	nr	0.27	2.58	4.21	1.02	7.81
25 x 3 Lead covered copper	nr	0.20	1.91	8.54	1.57	12.02
20 x 3 Bare aluminium	nr	0.20	1.91	1.03	0.44	3.38

EARTHING AND BONDING

	Unit	Labour hours	Net labour (£)	Net material (£)	O'heads /profit (£)	Total (£)
25 x 3 Bare aluminium	nr	0.20	1.91	0.90	0.42	3.23
25 x 6 Bare aluminium	nr	0.20	1.91	3.13	0.76	5.80
50 x 6 Bare aluminium	nr	0.27	2.58	3.34	0.89	6.81
25 x 3 PVC covered aluminium	nr	0.20	1.91	1.83	0.56	4.30

Tape clip

Size mm   Conductor material

	Unit	Labour hours	Net labour (£)	Net material (£)	O'heads /profit (£)	Total (£)
20 x 3 aluminium	nr	0.20	1.91	1.01	0.44	2.92
25 x 3 aluminium	nr	0.20	1.91	0.31	0.33	2.22
20 x 3 copper	nr	0.20	1.91	0.31	0.33	2.22
25 x 3 copper	nr	0.20	1.91	0.31	0.33	2.22
25 x 3 PVC covered copper	nr	0.20	1.91	0.75	0.40	3.06

Test and junction clamps

Square tape clamp

	Unit	Labour hours	Net labour (£)	Net material (£)	O'heads /profit (£)	Total (£)
25 x 3mm copper	nr	0.20	1.91	2.71	0.69	5.31
25 x 6mm copper	nr	0.20	1.91	11.89	2.07	13.80
50 x 6mm copper	nr	0.33	3.16	12.14	2.29	15.30
25 x 3mm aluminium	nr	0.20	1.91	2.41	0.65	4.32

Oblong test or junction clamp

	Unit	Labour hours	Net labour (£)	Net material (£)	O'heads /profit (£)	Total (£)
26 x 8mm copper	nr	0.20	1.91	4.20	0.92	7.03
26 x 8mm aluminium	nr	0.20	1.91	4.36	0.94	7.21

Plate type test clamp

	Unit	Labour hours	Net labour (£)	Net material (£)	O'heads /profit (£)	Total (£)
26 x 12mm	nr	0.33	3.16	11.56	2.21	16.93

Screw down test

	Unit	Labour hours	Net labour (£)	Net material (£)	O'heads /profit (£)	Total (£)
26 x 8mm	nr	0.30	2.87	10.40	1.99	15.26

Bimetallic connector

	Unit	Labour hours	Net labour (£)	Net material (£)	O'heads /profit (£)	Total (£)
26 x 4mm	nr	0.33	3.16	18.29	3.22	24.67

COMMUNICATIONS

	Unit	Labour hours	Net labour (£)	Net material (£)	O'heads /profit (£)	Total (£)
Square clamp						
8mm diameter copper	nr	0.27	2.58	3.19	0.87	6.64
8mm diameter aluminium	nr	0.27	2.58	2.61	0.78	5.97
Tee clamp						
8mm diameter copper	nr	0.30	2.87	3.19	0.91	6.97
8mm diameter aluminium	nr	0.30	2.87	2.61	0.82	6.30
Jointing clamp						
8mm diameter copper	nr	0.30	2.87	2.95	0.87	6.69
8mm diameter aluminium	nr	0.30	2.87	2.61	0.82	6.30
Test clamp						
8mm diameter copper	nr	0.27	2.58	2.95	0.83	6.36
8mm diameter aluminium	nr	0.27	2.58	2.36	0.74	4.94
Metalwork bond						
8mm diameter copper	nr	0.33	3.16	6.48	1.45	11.09
8mm diameter aluminium	nr	0.33	3.16	6.21	1.41	10.78
Pipebond						
pipe diameter 8mm, 50-200mm	nr	0.30	2.87	13.26	2.42	16.13
pipe diameter 8mm, 50-200mm	nr	0.30	2.87	12.11	2.25	17.23
Re-bar clamp, copper conductor						
pipe diameter 8mm, 8-18mm	nr	0.30	2.87	7.21	1.51	11.59
pipe diameter 8mm, 18-38mm	nr	0.30	2.87	19.84	3.41	26.12

Earth rods - copper bond rods

Diameter mm	Length mm	Unit	Labour hours	Net labour (£)	Net material (£)	O'heads /profit (£)	Total (£)
9.5	1200	nr	0.28	2.68	3.66	0.95	6.34
12.5	1200	nr	0.28	2.68	5.02	1.16	7.70
12.5	1500	nr	0.28	2.68	6.23	1.34	8.91
12.5	1800	nr	0.30	2.87	7.39	1.54	10.26

EARTHING AND BONDING

Earth rods (cont'd)		Unit	Labour hours	Net labour (£)	Net material (£)	O'heads /profit (£)	Total (£)
16	1200	nr	0.28	2.68	5.54	1.23	8.22
16	1500	nr	0.28	2.68	6.76	1.42	9.44
16	1800	nr	0.30	2.87	8.15	1.65	11.02
16	2400	nr	0.33	3.16	10.92	2.11	14.08
16	3000	nr	0.35	3.35	13.56	2.54	16.91
20	1200	nr	0.28	2.68	7.11	1.47	9.79
20	1500	nr	0.28	2.68	8.82	1.73	11.50
20	1800	nr	0.30	2.87	10.53	2.01	13.40
20	2400	nr	0.33	3.16	14.01	2.58	17.17
20	3000	nr	0.35	3.35	17.47	3.12	20.82

Fittings

	Unit	Labour hours	Net labour (£)	Net material (£)	O'heads /profit (£)	Total (£)
12.5mm coupling	nr	0.07	0.67	2.34	0.45	3.46
16mm coupling	nr	0.07	0.67	1.83	0.38	2.88
20mm coupling	nr	0.07	0.67	2.78	0.52	3.97
12mm driving stud	nr	0.07	0.67	1.72	0.36	2.75
16mm driving stud	nr	0.07	0.67	0.37	0.16	1.20
20mm driving stud	nr	0.07	0.67	11.97	1.90	14.54

Solid copper rod
Diameter Length
 mm   mm

		Unit	Labour hours	Net labour (£)	Net material (£)	O'heads /profit (£)	Total (£)
15	1200	nr	0.28	2.68	11.04	2.06	15.78
20	1200	nr	0.28	2.68	19.42	3.31	25.41

Fittings

	Unit	Labour hours	Net labour (£)	Net material (£)	O'heads /profit (£)	Total (£)
15mm diameter driving stud	nr	0.07	0.67	0.73	0.21	1.61
20mm diameter driving stud	nr	0.07	0.67	1.20	0.28	2.15
coupling dowel	nr	0.07	0.67	0.59	0.19	1.45
15mm diameter spike	nr	0.07	0.67	0.69	0.20	1.56
20mm diameter spike	nr	0.07	0.67	1.11	0.27	1.78

Stainless steel rod

	Unit	Labour hours	Net labour (£)	Net material (£)	O'heads /profit (£)	Total (£)
16mm diameter x 1200mm long	nr	0.28	2.68	21.96	3.70	28.34

COMMUNICATIONS

	Unit	Labour hours	Net labour (£)	Net material (£)	O'heads /profit (£)	Total (£)

Earth rod clamps

Rod to tape clamp

Rod diameter mm	Conductor size mm					
12.5	26 x 12	nr	0. 18	1. 72	2. 16	0. 58	4. 46
16	26 x 12	nr	0. 18	1. 72	2. 16	0. 58	4. 46
20	26 x 10	nr	0. 18	1. 72	2. 16	0. 58	4. 46
16	40 x 12	nr	0. 18	1. 72	5. 98	1. 16	8. 86
16	51 x 8	nr	0. 18	1. 72	8. 15	1. 48	11. 35
20	51 x 12	nr	0. 18	1. 72	8. 58	1. 54	11. 84
12.5	26 x 20	nr	0. 18	1. 72	8. 58	1. 54	11. 84
16	26 x 18	nr	0. 18	1. 72	6. 42	1. 22	9. 36
20	26 x 10	nr	0. 18	1. 72	6. 42	1. 22	9. 36
25	26 x 10	nr	0. 18	1. 72	6. 42	1. 22	9. 36

Rod to cable clamp

Rod diameter mm	Conductor size mm					
9.5	6-35	nr	0. 25	2. 39	1. 08	0. 52	3. 99
12.5	16-50	nr	0. 25	2. 39	1. 49	0. 58	4. 46
16	16-70	nr	0. 25	2. 39	1. 27	0. 55	4. 21
20	16-95	nr	0. 25	2. 39	1. 43	0. 57	4. 39
25	70-120	nr	0. 25	2. 39	2. 95	0. 80	6. 14

'U' bolt clamp, rod diameter

16mm		nr	0. 25	2. 39	3. 20	0. 84	6. 43
20mm		nr	0. 25	2. 39	3. 65	0. 91	6. 95
25mm		nr	0. 25	2. 39	4. 47	1. 03	7. 89

Rod to cable clamp ('U' bolt type)

Rod diameter mm	Conductor size mm2					
16 or 20	16-70	nr	0. 25	2. 39	5. 30	1. 15	8. 84
16 or 20	70-150	nr	0. 25	2. 39	5. 45	1. 18	9. 02

EARTHING AND BONDING

		Unit	Labour hours	Net labour (£)	Net material (£)	O'heads /profit (£)	Total (£)
Rod to cable lug clamp							
Rod diameter mm	Rod type						
9.5	copper bond	nr	0.25	2.39	2.23	0.69	5.31
16	copper bond	nr	0.25	2.39	11.64	2.10	16.13
16	solid copper	nr	0.25	2.39	11.80	2.13	16.32
20	copper bond	nr	0.25	2.39	11.64	2.10	16.13
20	solid copper	nr	0.25	2.39	11.80	2.13	16.32
Bonds and clamps							
B bond							
copper		nr	0.28	2.68	1.24	0.59	4.51
aluminium		nr	0.28	2.68	1.48	0.62	4.78
RWP bond		nr	0.00	0.00	3.00	0.45	3.00
copper		nr	0.28	2.68	3.00	0.85	6.53
aluminium		nr	0.28	2.68	2.65	0.80	6.13
Water main bond							
copper		nr	0.28	2.68	4.19	1.03	7.90
Flexible copper earth bond fixed to backgrounds with screws							
25 x 3mm with 200mm centre holes		nr	0.23	2.20	5.70	1.19	9.09
25 x 3mm with 400mm centre holes		nr	0.23	2.20	8.20	1.56	11.96
Earth points fixed to backgrounds							
four hole		nr	0.20	1.91	13.04	2.24	17.19
2 hole with 25 x 3mm/70mm2 front plate		nr	0.20	1.91	9.19	1.66	12.76
2 hole with 25 x 3mm/8mm diameter front plate		nr	0.20	1.91	9.19	1.66	12.76
2 hole without front plate		nr	0.20	1.91	8.22	1.52	11.65

COMMUNICATIONS

	Unit	Labour hours	Net labour (£)	Net material (£)	O'heads /profit (£)	Total (£)
Earth points with welded tails, fixed to backgrounds						
four hole with 0.5m long tail of 70mm2 PVC/copper cable	nr	0. 20	1. 91	20. 06	3. 30	21. 97
2 hole with 25 x 3mm/70mm2 front plate with 0.5mm long tail of 70mm2 PVC/copper cable	nr	0. 20	1. 91	15. 49	2. 61	20. 01

Y

MECHANICAL AND ELECTRICAL MEASUREMENT SERVICES

	Unit	Labour hours	Net labour (£)	Net material (£)	O'heads /profit (£)	Total (£)
Y41 FANS						
General notes						

1. Prices do not include any builders' work involved in fitting the ventilation fans

2. Fused spur unit and cable measured separately

Window type, fixing in prepared opening

	Unit	Labour hours	Net labour (£)	Net material (£)	O'heads /profit (£)	Total (£)
standard fan	nr	1.50	14.36	28.15	6.38	48.89
with timer	nr	1.50	14.36	36.85	7.68	58.89
with shutter	nr	1.50	14.36	36.85	7.68	58.89
with timer and shutter	nr	1.50	14.36	45.89	9.04	69.29
with humidistat	nr	1.50	14.36	59.61	11.10	85.07
with humidistat and shutter	nr	1.50	14.36	69.14	12.53	96.03

Wall type, fixing in prepared opening

	Unit	Labour hours	Net labour (£)	Net material (£)	O'heads /profit (£)	Total (£)
standard fan	nr	1.50	14.36	28.15	6.38	48.89
with timer	nr	1.50	14.36	36.85	7.68	58.89
with shutter	nr	1.50	14.36	36.85	7.68	58.89
with timer and shutter	nr	1.50	14.36	45.89	9.04	69.29
with humidistat	nr	1.50	14.36	59.61	11.10	85.07
with humidistat and shutter	nr	1.50	14.36	69.14	12.53	96.03

Panel type, fixing in prepared opening

	Unit	Labour hours	Net labour (£)	Net material (£)	O'heads /profit (£)	Total (£)
standard fan	nr	1.50	14.36	21.09	5.32	40.77
with timer	nr	1.50	14.36	29.79	6.62	50.77
with shutter	nr	1.50	14.36	29.79	6.62	50.77
with timer and shutter	nr	1.50	14.36	32.82	7.08	54.26
with humidistat	nr	1.50	14.36	52.54	10.04	76.94
with humidistat and shutter	nr	1.50	14.36	62.08	11.47	87.91

FANS

	Unit	Labour hours	Net labour (£)	Net material (£)	O'heads /profit (£)	Total (£)
Domestic fan controllers						
multi-unit controller (up to 5 units); surface fixed to backgrounds	nr	0.50	4.79	55.83	9.09	69.71
multi-unit controller (up to 5 units); flush mounted with fixings	nr	0.50	4.79	59.54	9.65	73.98
air quality controller; surface fixed to backgrounds	nr	0.50	4.79	73.54	11.75	90.08
Controller range extractor fans, complete with electronic shutter						
Window type, fixing in prepared opening						
6" diameter	nr	1.50	14.36	94.56	16.34	125.26
7" diameter	nr	1.50	14.36	123.79	20.72	158.87
9" diameter	nr	1.50	14.36	161.73	26.41	202.50
12" diameter	nr	1.83	17.51	224.63	36.32	278.46
Roof type, fixing in prepared opening						
6" diameter	nr	3.00	28.71	104.38	19.96	153.05
7" diameter	nr	3.00	28.71	144.52	25.98	199.21
9" diameter	nr	3.00	28.71	204.79	35.02	268.52
12" diameter	nr	3.50	33.50	252.53	42.90	328.93
Wall type, fixing in prepared opening						
6" diameter	nr	2.00	19.14	119.60	20.81	159.55
7" diameter	nr	2.00	19.14	161.37	28.51	180.51
9" diameter	nr	2.00	19.14	204.79	33.59	257.52
12" diameter	nr	2.50	23.93	287.75	46.75	358.43
Panel type, fixing in prepared opening						
6" diameter	nr	2.00	19.14	103.61	18.41	141.16
7" diameter	nr	2.00	19.14	132.53	22.75	174.42

	Unit	Labour hours	Net labour (£)	Net material (£)	O'heads /profit (£)	Total (£)
9" diameter	nr	2.00	19.14	170.86	28.50	218.50
12" diameter	nr	2.50	23.93	234.26	38.73	296.92
Darkroom/X-ray type, fixing to prepared opening						
6" diameter	nr	2.00	19.14	117.08	20.43	156.65
7" diameter	nr	2.00	19.14	166.79	27.89	213.82
9" diameter	nr	2.00	19.14	205.22	33.65	258.01
12" diameter	nr	2.50	23.93	267.77	43.75	335.45
In-line type, fixing to ductwork						
175mm diameter duct	nr	2.00	19.14	107.33	18.97	145.44
225mm diameter duct	nr	2.00	19.14	148.53	25.15	192.82
300mm diameter duct	nr	2.00	19.14	187.43	30.99	237.56
400mm diameter duct	nr	2.50	23.93	259.55	42.52	326.00
Commercial range extractor fans without shutters						
Window type, fixing in prepared opening						
6" diameter	nr	1.50	14.36	72.95	13.10	100.41
7" diameter	nr	1.50	14.36	96.86	16.68	127.90
9" diameter	nr	1.50	14.36	126.80	21.17	162.33
12" diameter	nr	1.83	17.51	178.38	29.38	225.27
Roof type, fixing in prepared opening						
6" diameter	nr	3.00	28.71	82.92	16.74	128.37
7" diameter	nr	3.00	28.71	117.59	21.95	168.25
9" diameter	nr	3.00	28.71	148.19	26.54	203.44
12" diameter	nr	3.50	33.50	205.78	35.89	275.17
Wall type, fixing in prepared opening						
6" diameter	nr	2.00	19.14	98.14	19.03	117.28
7" diameter	nr	2.00	19.14	135.58	23.21	177.93
9" diameter	nr	2.00	19.14	169.80	28.34	217.28
12" diameter	nr	2.50	23.93	222.85	37.02	283.80

FANS

	Unit	Labour hours	Net labour (£)	Net material (£)	O'heads /profit (£)	Total (£)
Panel type, fixing in prepared opening						
6" diameter	nr	2.00	19.14	82.14	15.19	116.4
7" diameter	nr	2.00	19.14	105.60	18.71	143.4
9" diameter	nr	2.00	19.14	135.94	23.26	178.3
12" diameter	nr	2.50	23.93	186.87	31.62	242.4
Darkroom/X-ray type, fixing in prepared opening						
6" diameter	nr	2.00	19.14	95.26	17.16	114.4
7" diameter	nr	2.00	19.14	166.79	27.89	213.8
9" diameter	nr	2.00	19.14	205.22	33.65	258.0
12" diameter	nr	2.50	23.93	267.77	46.75	291.7
In-line type, fixing to ductwork						
6" diameter	nr	2.00	19.14	85.23	15.66	120.0
7" diameter	nr	2.00	19.14	120.90	21.01	161.0
9" diameter	nr	2.00	19.14	151.52	25.60	196.2
12" diameter	nr	2.50	23.93	211.45	35.31	270.6
Commercial fan controllers						
individual controller, surface mounted, fixed to backgrounds	nr	0.50	4.79	33.92	6.11	38.7
individual fan controller, flush mounted, fixed to backgrounds	nr	0.50	4.79	37.12	6.29	48.2
air quality controller, surface mounted fixed to backgrounds	nr	0.50	4.79	73.54	11.75	78.33

	Unit	Labour hours	Net labour (£)	Net material (£)	O'heads /profit (£)	Total (£)

Y60 CONDUIT AND CABLE TRUNKING

General notes

1. All labour times include for installing, cutting, threading and bending conduit

2. All labour times include installing materials up to maximum height of 4.5m, but not erecting tressles or scaffolding. Add 10% to times for heights between 4.5m and 6.5m

3. Allow additional time for breaking into existing conduit

4. A discount of 65% has been incorporated in the nett material costs for steel conduit and 45% for PVC conduit. Separate discounts have been used for accessories, stated eleswhere

5. A waste factor of 10% has been incorporated in all the conduit length costs

6. Steel - galvanized finish to BS4568 class 4 and BS31 class 4

7. Black enamel finish to BS4568 class 2 and BS31 class 2

8. PVC manufactured to BS4607 and BS6099

CONDUIT AND CABLE TRUNKING

	Unit	Labour hours	Net labour (£)	Net material (£)	O'heads /profit (£)	Total (£)
Steel conduit, light gauge, galvanized						
Run on surface						
16mm	m	0.22	2.11	1.30	0.51	3.92
20mm	m	0.29	2.78	1.30	0.61	4.69
25mm	m	0.37	3.54	1.77	0.80	6.11
Run in wall chase or floor screed						
16mm	m	0.18	1.72	1.30	0.45	3.47
20mm	m	0.25	2.39	1.30	0.55	4.24
25mm	m	0.33	3.16	1.77	0.74	5.67
Run in floor slab						
16mm	m	0.17	1.63	1.30	0.44	3.37
20mm	m	0.20	1.91	1.30	0.48	3.69
25mm	m	0.25	2.39	1.77	0.62	4.78
Steel conduit, light gauge, black enamel						
Run on surface						
16mm	m	0.22	2.11	0.98	0.46	3.55
20mm	m	0.29	2.78	0.98	0.56	4.32
25mm	m	0.37	3.54	1.13	0.70	5.37
Run in wall chase or floor screed						
16mm	m	0.18	1.72	0.98	0.40	3.10
20mm	m	0.25	2.39	0.98	0.51	3.88
25mm	m	0.33	3.16	1.13	0.64	4.93
Run in floor slab						
16mm	m	0.17	1.63	0.98	0.39	3.00
20mm	m	0.20	1.91	0.98	0.43	3.32
25mm	m	0.25	2.39	1.13	0.53	4.05

	Unit	Labour hours	Net labour (£)	Net material (£)	O'heads /profit (£)	Total (£)
Steel conduit, heavy gauge, galvanized						
Run on surface						
16mm	m	0.28	2.68	1.29	0.60	4.57
20mm	m	0.33	3.16	1.29	0.67	5.12
25mm	m	0.42	4.02	1.77	0.87	6.66
32mm	m	0.48	4.59	2.11	1.00	7.70
Run in wall chase or floor screed						
16mm	m	0.25	2.39	1.29	0.55	4.23
20mm	m	0.32	3.06	1.29	0.65	5.00
25mm	m	0.35	3.35	1.77	0.77	5.89
32mm	m	0.43	4.12	2.11	0.93	7.16
Run in floor slab						
16mm	m	0.23	2.20	1.29	0.52	4.01
20mm	m	0.25	2.39	1.29	0.55	4.23
25mm	m	0.33	3.16	1.77	0.74	5.67
32mm	m	0.42	4.02	2.11	0.92	7.05
Steel conduit, heavy gauge, black enamel						
Run on surface						
16mm	m	0.28	2.68	0.98	0.55	4.21
20mm	m	0.33	3.16	0.98	0.62	4.76
25mm	m	0.42	4.02	1.13	0.77	5.92
32mm	m	0.48	4.59	1.50	0.91	7.00
Run in wall chase or floor screed						
16mm	m	0.25	2.39	0.98	0.51	3.88
20mm	m	0.32	3.06	0.98	0.61	4.65
25mm	m	0.35	3.35	1.13	0.67	5.15
32mm	m	0.43	4.12	1.50	0.84	6.46

CONDUIT AND CABLE TRUNKING

	Unit	Labour hours	Net labour (£)	Net material (£)	O'heads /profit (£)	Total (£)
Run in floor slab						
16mm	m	0.23	2.20	0.98	0.48	3.66
20mm	m	0.25	2.39	0.98	0.51	3.88
25mm	m	0.33	3.16	1.13	0.64	4.93
32mm	m	0.42	4.02	1.50	0.83	6.35
PVC conduit, super high impact, light gauge						
Run on surface						
16mm	m	0.17	1.63	0.49	0.32	2.44
20mm	m	0.20	1.91	0.62	0.38	2.91
25mm	m	0.62	5.93	0.62	0.98	7.53
32mm	m	0.30	2.87	1.37	0.64	4.88
Run in wall chase or floor screed						
16mm	m	0.15	1.44	0.49	0.29	2.22
20mm	m	0.17	1.63	0.62	0.34	2.59
25mm	m	0.22	2.11	0.62	0.41	3.14
32mm	m	0.28	2.68	1.37	0.61	4.66
Run in floor slab						
16mm	m	0.13	1.24	0.49	0.26	1.99
20mm	m	0.15	1.44	0.62	0.31	2.37
25mm	m	0.20	1.91	0.62	0.38	2.91
32mm	m	0.25	2.39	1.37	0.56	4.32
PVC conduit, super high impact, heavy gauge						
Run on surface						
16mm	m	0.17	1.63	0.76	0.36	2.75
20mm	m	0.20	1.91	0.90	0.42	3.23
25mm	m	0.25	2.39	1.22	0.54	4.15
32mm	m	0.30	2.87	1.96	0.72	5.55

M & E MEASUREMENT SERVICES

	Unit	Labour hours	Net labour (£)	Net material (£)	O'heads /profit (£)	Total (£)
Run in wall chase or floor screed						
16mm	m	0.15	1.44	0.76	0.33	2.53
20mm	m	0.17	1.63	0.90	0.38	2.91
25mm	m	0.22	2.11	1.22	0.50	3.83
32mm	m	0.28	2.68	1.96	0.70	5.34
Run in floor slab						
16mm	m	0.13	1.24	0.76	0.30	2.30
20mm	m	0.15	1.44	0.90	0.35	2.69
25mm	m	0.20	1.91	1.22	0.47	3.60
32mm	m	0.25	2.39	1.96	0.65	5.00
Flexible PVC conduit						
Run on surface						
16mm	m	0.15	1.44	1.21	0.40	3.05
20mm	m	0.18	1.72	1.31	0.45	3.48
25mm	m	0.22	2.11	1.62	0.56	4.29
32mm	m	0.27	2.58	2.15	0.71	5.44
Run in floor slab						
16mm	m	0.08	0.77	1.21	0.30	2.28
20mm	m	0.08	0.77	1.31	0.31	2.39
25mm	m	0.08	0.77	1.62	0.36	2.75
32mm	m	0.08	0.77	2.15	0.44	3.36

121

CONDUIT AND CABLE TRUNKING

	Unit	Labour hours	Net labour (£)	Net material (£)	O'heads /profit (£)	Total (£)
Steel conduit fittings						
A discount of 57% has been incorporated within the material costs for steel fittings						
Steel fittings for connection to conduit including cutting and threading conduit						
Bends, galvanized						
Solid, internal thread						
20mm	nr	0.07	0.67	0.99	0.25	1.91
25mm	nr	0.08	0.77	1.46	0.33	2.56
32mm	nr	0.08	0.77	2.89	0.55	4.21
Inspection type						
20mm	nr	0.10	0.96	1.50	0.37	2.83
25mm	nr	0.12	1.15	2.35	0.53	4.03
32mm	nr	0.12	1.15	5.56	1.01	7.72
Bends, black enamel						
Solid, internal thread						
20mm	nr	0.07	0.67	0.78	0.22	1.67
25mm	nr	0.08	0.77	1.23	0.30	2.30
32mm	nr	0.08	0.77	2.06	0.42	3.25
Inspection type						
20mm	nr	0.10	0.96	1.26	0.33	2.55
25mm	nr	0.12	1.15	1.94	0.46	3.55
32mm	nr	0.12	1.15	4.19	0.80	6.14

	Unit	Labour hours	Net labour (£)	Net material (£)	O'heads /profit (£)	Total (£)
Tees, galvanized						
Inspection type						
20mm	nr	0.15	1.44	1.67	0.47	3.58
25mm	nr	0.17	1.63	2.21	0.58	4.42
32mm	nr	0.17	1.63	6.20	1.17	9.00
Tees, black enamel						
Inspection type						
20mm	nr	0.15	1.44	1.38	0.42	3.24
25mm	nr	0.17	1.63	1.83	0.52	3.98
32mm	nr	0.17	1.63	4.80	0.96	7.39
Elbows, galvanized						
Inspection type						
20mm	nr	0.10	0.96	1.44	0.36	2.76
25mm	nr	0.12	1.15	1.69	0.43	3.27
32mm	nr	0.12	1.15	3.42	0.69	5.26
Elbows, black enamel						
Inspection type						
20mm	nr	0.10	0.96	1.18	0.32	2.46
25mm	nr	0.12	1.15	1.41	0.38	2.94
32mm	nr	0.12	1.15	2.72	0.58	4.45

Adaptable galvanized steel boxes

Depth	Size	Unit	Labour hours	Net labour (£)	Net material (£)	O'heads /profit (£)	Total (£)
37.5	75 x 75mm	nr	0.20	1.91	2.43	0.65	4.34
37.5	100 x 100mm	nr	0.20	1.91	2.60	0.68	4.51
37.5	150 x 75mm	nr	0.20	1.91	2.61	0.68	4.52
50	75 x 75mm	nr	0.20	1.91	2.59	0.67	4.50
50	100 x 100mm	nr	0.20	1.91	2.89	0.72	4.80
50	150 x 75mm	nr	0.20	1.91	3.01	0.74	4.92
50	150 x 100mm	nr	0.20	1.91	3.02	0.74	4.93
50	150 x 150mm	nr	0.20	1.91	3.77	0.85	5.68

CONDUIT AND CABLE TRUNKING

Conduit fittings (cont'd)		Unit	Labour hours	Net labour (£)	Net material (£)	O'heads /profit (£)	Total (£)
75	225 x 150mm	nr	0.33	3.16	6.76	1.49	9.92
75	225 x 225mm	nr	0.33	3.16	8.59	1.76	11.75
75	300 x 150mm	nr	0.33	3.16	8.55	1.76	11.71
75	300 x 300mm	nr	0.33	3.16	11.05	2.13	14.21
100	150 x 150mm	nr	0.20	1.91	5.74	1.15	7.65
100	225 x 225mm	nr	0.33	3.16	9.35	1.88	12.51

Adaptable black enamel steel boxes

Depth	Size						
37.5	75 x 75mm	nr	0.20	1.91	2.06	0.60	3.97
37.5	100 x 100mm	nr	0.20	1.91	2.25	0.62	4.16
37.5	150 x 75mm	nr	0.20	1.91	2.30	0.63	4.21
50	75 x 75mm	nr	0.20	1.91	2.18	0.61	4.09
50	100 x 100mm	nr	0.20	1.91	2.37	0.64	4.28
50	150 x 75mm	nr	0.20	1.91	2.41	0.65	4.32
50	150 x 100mm	nr	0.20	1.91	2.71	0.69	4.62
50	150 x 150mm	nr	0.20	1.91	3.17	0.76	5.08
75	150 x 75mm	nr	0.20	1.91	2.83	0.71	4.74
75	150 x 150mm	nr	0.20	1.91	3.51	0.81	5.42
75	225 x 150mm	nr	0.33	3.16	5.23	1.26	8.39
75	225 x 225mm	nr	0.33	3.16	6.42	1.44	9.58
75	300 x 150mm	nr	0.33	3.16	6.39	1.43	9.55
75	300 x 300mm	nr	0.33	3.16	9.44	1.89	12.60
100	150 x 150mm	nr	0.20	1.91	4.39	0.94	6.30
100	225 x 225mm	nr	0.33	3.16	7.00	1.52	10.16

Steel boxes, circular type, galvanized

Terminal

20mm		nr	0.09	0.86	1.29	0.32	2.47
25mm		nr	0.09	0.86	1.79	0.40	3.05

Back outlet

20mm		nr	0.09	0.86	2.12	0.45	3.43

M & E MEASUREMENT SERVICES

	Unit	Labour hours	Net labour (£)	Net material (£)	O'heads /profit (£)	Total (£)
Terminal and back outlet						
20mm	nr	0.09	0.86	2.39	0.49	3.74
Through way						
20mm	nr	0.09	0.86	1.47	0.35	2.68
25mm	nr	0.09	0.86	2.13	0.45	3.44
Through and back outlet						
20mm	nr	0.09	0.86	2.65	0.53	4.04
Angle						
20mm	nr	0.09	0.86	1.47	0.35	2.68
25mm	nr	0.09	0.86	2.13	0.45	3.44
Tee						
20mm	nr	0.09	0.86	1.64	0.37	2.87
25mm	nr	0.09	0.86	2.27	0.47	3.60
Angle tangent						
20mm	nr	0.09	0.86	2.50	0.50	3.86
Tee tangent						
20mm	nr	0.09	0.86	2.64	0.53	4.03
Branch 'U'						
20mm	nr	0.09	0.86	2.78	0.55	4.19
Branch tee						
20mm	nr	0.09	0.86	2.78	0.55	4.19
Four way						
20mm	nr	0.09	0.86	1.92	0.42	3.20
25mm	nr	0.09	0.86	2.66	0.53	4.05

CONDUIT AND CABLE TRUNKING

	Unit	Labour hours	Net labour (£)	Net material (£)	O'heads /profit (£)	Total (£)
Boxes, circular type, deep pattern, galvanized						
Terminal - 20mm	nr	0.09	0.86	5.99	1.03	7.88
Through way - 20mm	nr	0.09	0.86	6.38	1.09	8.33
Angle, 20mm	nr	0.09	0.86	6.38	1.09	8.33
Tee - 20mm	nr	0.17	1.63	6.70	1.25	8.33
Four way	nr	0.17	1.63	6.99	1.29	9.91
Boxes, extensions galvanized						
Terminal - 20mm	nr	0.09	0.86	3.99	0.73	5.58
Through way - 20mm	nr	0.09	0.86	4.38	0.79	6.03
Boxes, large circular, galvanized						
Terminal						
20mm	nr	0.09	0.86	7.23	1.21	9.30
25mm	nr	0.09	0.86	7.23	1.21	9.30
32mm	nr	0.09	0.86	7.23	1.21	9.30
1.5"	nr	0.09	0.86	9.24	1.51	11.61
Through						
20mm	nr	0.09	0.86	8.26	1.37	10.49
25mm	nr	0.09	0.86	8.26	1.37	10.49
32mm	nr	0.09	0.86	8.26	1.37	10.49
1.5"	nr	0.09	0.86	10.66	1.73	13.25
Angle						
20mm	nr	0.09	0.86	8.26	1.37	10.49
25mm	nr	0.09	0.86	8.26	1.37	10.49
32mm	nr	0.09	0.86	8.26	1.37	10.49
1.5"	nr	0.09	0.86	10.66	1.73	13.25

	Unit	Labour hours	Net labour (£)	Net material (£)	O'heads /profit (£)	Total (£)
Tee						
20mm	nr	0.09	0.86	11.26	1.82	13.94
25mm	nr	0.09	0.86	11.26	1.82	13.94
32mm	nr	0.09	0.86	11.26	1.82	13.94
1.5"	nr	0.09	0.86	12.66	2.03	15.55
Four way						
20mm	nr	0.09	0.86	11.59	1.87	14.32
25mm	nr	0.09	0.86	11.59	1.87	14.32
32mm	nr	0.09	0.86	11.59	1.87	14.32
1.5"	nr	0.09	0.86	15.97	2.52	19.35
Boxes, rectangular junction, galvanized						
Through way						
20mm	nr	0.09	0.86	6.05	1.04	7.95
25mm	nr	0.09	0.86	7.15	1.20	9.21
32mm	nr	0.09	0.86	14.76	2.34	17.96
1.5"	nr	0.09	0.86	17.85	2.81	21.52
Tee						
20mm	nr	0.09	0.86	6.63	1.12	8.61
25mm	nr	0.09	0.86	8.69	1.43	10.98
32mm	nr	0.09	0.86	16.08	2.54	19.48
Four way						
20mm	nr	0.09	0.86	7.65	1.28	9.79
25mm	nr	0.09	0.86	9.43	1.54	11.83
32mm	nr	0.09	0.86	16.75	2.64	20.25
Conduit fittings for steel boxes, circular type, black enamel						
Terminal						
20mm	nr	0.09	0.86	1.06	0.29	2.21
25mm	nr	0.09	0.86	1.47	0.35	2.68

CONDUIT AND CABLE TRUNKING

Conduit fittings (cont'd)	Unit	Labour hours	Net labour (£)	Net material (£)	O'heads /profit (£)	Total (£)
Back outlet						
20mm	nr	0. 09	0. 86	1. 75	0. 39	3. 00
Terminal and back outlet						
20mm	nr	0. 09	0. 86	1. 98	0. 43	3. 27
Through way						
20mm	nr	0. 09	0. 86	1. 26	0. 32	2. 44
25mm	nr	0. 09	0. 86	1. 75	0. 39	3. 00
Through way and back outlet						
20mm	nr	0. 09	0. 86	2. 21	0. 46	3. 53
Angle						
20mm	nr	0. 09	0. 86	1. 26	0. 32	2. 44
25mm	nr	0. 09	0. 86	1. 75	0. 39	3. 00
Tee						
20mm	nr	0. 09	0. 86	1. 35	0. 33	2. 54
25mm	nr	0. 09	0. 86	1. 88	0. 41	3. 15
Angle tangent						
20mm	nr	0. 09	0. 86	2. 06	0. 44	3. 36
Tee tangent						
20mm	nr	0. 09	0. 86	2. 64	0. 53	4. 03
Branch 'U'						
20mm	nr	0. 09	0. 86	1. 84	0. 40	3. 10
Branch tee						
20mm	nr	0. 09	0. 86	2. 29	0. 47	3. 62

M & E MEASUREMENT SERVICES

	Unit	Labour hours	Net labour (£)	Net material (£)	O'heads /profit (£)	Total (£)
Four way						
20mm	nr	0.09	0.86	1.58	0.37	2.81
25mm	nr	0.09	0.86	2.19	0.46	3.51
Boxes, circular type, deep pattern, black enamel						
Terminal - 20mm	nr	0.09	0.86	4.64	0.82	6.32
Through way - 20mm	nr	0.09	0.86	4.95	0.87	6.68
Angle - 20mm	nr	0.09	0.86	4.95	0.87	6.68
Tee - 20mm	nr	0.09	0.86	5.19	0.91	6.96
Four way - 20mm	nr	0.09	0.86	5.42	0.94	7.22
Boxes, extensions, black enamel						
Terminal - 20mm	nr	0.09	0.86	3.24	0.61	4.71
Through way - 20mm	nr	0.09	0.86	3.47	0.65	4.98
Boxes, large circular, black enamel						
Terminal						
20mm	nr	0.09	0.86	5.61	0.97	7.44
25mm	nr	0.09	0.86	5.61	0.97	7.44
32mm	nr	0.09	0.86	5.61	0.97	7.44
1.5"	nr	0.09	0.86	6.93	1.17	8.96
Through way						
20mm	nr	0.09	0.86	6.42	1.09	8.37
25mm	nr	0.09	0.86	6.42	1.09	8.37
32mm	nr	0.09	0.86	6.42	1.09	8.37
1.5"	nr	0.09	0.86	8.00	1.33	10.19

CONDUIT AND CABLE TRUNKING

Conduit fittings (cont'd)	Unit	Labour hours	Net labour (£)	Net material (£)	O'heads /profit (£)	Total (£)
Angle						
20mm	nr	0.09	0.86	6.42	1.09	8.37
25mm	nr	0.09	0.86	6.42	1.09	8.37
32mm	nr	0.09	0.86	6.42	1.09	8.37
1.5"	nr	0.09	0.86	8.00	1.33	10.19
Tee						
20mm	nr	0.09	0.86	8.75	1.44	11.05
25mm	nr	0.09	0.86	8.75	1.44	11.05
32mm	nr	0.09	0.86	8.75	1.44	11.05
1.5"	nr	0.09	0.86	9.69	1.58	12.13
Four way						
20mm	nr	0.09	0.86	9.34	1.53	11.73
25mm	nr	0.09	0.86	9.34	1.53	11.73
32mm	nr	0.09	0.86	9.34	1.53	11.73
1.5"	nr	0.09	0.86	12.61	2.02	15.49
Boxes, rectangular junction, black enamel						
Through way						
20mm	nr	0.09	0.86	4.69	0.83	6.38
25mm	nr	0.09	0.86	5.54	0.96	7.36
32mm	nr	0.09	0.86	11.45	1.85	14.16
1.5"	nr	0.09	0.86	14.19	2.26	17.31
Tee						
20mm	nr	0.09	0.86	5.14	0.90	6.90
25mm	nr	0.09	0.86	6.74	1.14	8.74
32mm	nr	0.09	0.86	12.47	2.00	15.33
Intersection						
20mm	nr	0.09	0.86	5.93	1.02	7.81
25mm	nr	0.09	0.86	7.31	1.23	9.40
32mm	nr	0.09	0.86	12.99	2.08	15.93

	Unit	Labour hours	Net labour (£)	Net material (£)	O'heads /profit (£)	Total (£)
Saddles fixed to backgrounds, requiring drilling, plugging and screwing						
Saddle, galvanized steel						
20mm	nr	0. 12	1. 15	0. 10	0. 19	1. 44
25mm	nr	0. 12	1. 15	0. 14	0. 19	1. 48
32mm	nr	0. 15	1. 44	0. 24	0. 25	1. 93
Distance saddle, galvanized iron						
20mm	nr	0. 13	1. 24	0. 85	0. 31	2. 40
25mm	nr	0. 13	1. 24	1. 11	0. 35	2. 70
32mm	nr	0. 17	1. 63	1. 84	0. 52	3. 99
Spacer bar saddle, galvanized steel						
20mm	nr	0. 13	1. 24	0. 25	0. 22	1. 71
25mm	nr	0. 13	1. 24	0. 30	0. 23	1. 77
32mm	nr	0. 17	1. 63	0. 77	0. 36	2. 76
Saddle, black enamel steel						
20mm	nr	0. 12	1. 15	0. 09	0. 19	1. 43
25mm	nr	0. 12	1. 15	0. 11	0. 19	1. 45
32mm	nr	0. 15	1. 44	0. 19	0. 24	1. 87
Distance saddle, black enamel iron						
20mm	nr	0. 13	1. 24	0. 64	0. 28	2. 16
25mm	nr	0. 13	1. 24	1. 11	0. 35	2. 70
32mm	nr	0. 17	1. 63	1. 36	0. 45	3. 44
Spacer bar saddle, black enamel steel						
20mm	nr	0. 13	1. 24	0. 25	0. 22	1. 71
25mm	nr	0. 13	1. 24	0. 30	0. 23	1. 77
32mm	nr	0. 17	1. 63	0. 77	0. 36	2. 76

CONDUIT AND CABLE TRUNKING

	Unit	Labour hours	Net labour (£)	Net material (£)	O'heads /profit (£)	Total (£)
Couplings, galvanized steel, including threading conduit and connecting						
Standard coupling						
20mm	nr	0.16	1.53	0.23	0.26	2.02
25mm	nr	0.20	1.91	0.28	0.33	2.52
32mm	nr	0.32	3.06	0.84	0.58	4.48
Earthing						
20mm	nr	0.16	1.53	1.50	0.45	3.48
25mm	nr	0.20	1.91	1.87	0.57	4.35
Inspection						
20mm	nr	0.16	1.53	2.96	0.67	5.16
25mm	nr	0.20	1.91	4.51	0.96	7.38
Flanged						
20mm	nr	0.30	2.87	1.94	0.72	5.53
25mm	nr	0.30	2.87	2.94	0.87	6.68
32mm	nr	0.30	2.87	6.57	1.42	10.86
Couplings, black enamel steel including threading conduit and connecting						
Standard coupling						
20mm	nr	0.16	1.53	0.19	0.26	1.98
25mm	nr	0.20	1.91	0.22	0.32	2.45
32mm	nr	0.32	3.06	0.64	0.55	4.25
Earthing						
20mm	nr	0.16	1.53	1.16	0.40	3.09
25mm	nr	0.20	1.91	1.44	0.50	3.85

	Unit	Labour hours	Net labour (£)	Net material (£)	O'heads /profit (£)	Total (£)
Inspection						
20mm	nr	0.16	1.53	2.29	0.57	4.39
25mm	nr	0.20	1.91	3.50	0.81	6.22
Flanged						
20mm	nr	0.30	2.87	1.59	0.67	5.13
25mm	nr	0.30	2.87	2.40	0.79	6.06
32mm	nr	0.30	2.87	5.23	1.22	9.32
Bushes, brass, connected to conduit						
Circular female, standard						
20mm	nr	0.02	0.19	0.16	0.05	0.40
25mm	nr	0.02	0.19	0.21	0.06	0.46
32mm	nr	0.03	0.29	0.37	0.10	0.76
Hexagon male, light						
20mm	nr	0.02	0.19	0.23	0.06	0.48
25mm	nr	0.02	0.19	0.36	0.08	0.63
32mm	nr	0.03	0.29	0.68	0.15	1.12
Hexagon male, standard						
20mm	nr	0.02	0.19	0.28	0.07	0.54
25mm	nr	0.02	0.19	0.51	0.10	0.80
32mm	nr	0.03	0.29	0.98	0.19	1.46
Adaptors (male to female) connected to conduit						
Galvanized, steel						
3/4" - 20mm	nr	0.03	0.29	0.65	0.14	1.08
1" - 25mm	nr	0.03	0.29	0.77	0.16	1.22
20mm - 3/4"	nr	0.03	0.29	0.65	0.14	1.08
25mm - 1"	nr	0.03	0.29	0.77	0.16	1.22

CONDUIT AND CABLE TRUNKING

Conduit fittings (cont'd)	Unit	Labour hours	Net labour (£)	Net material (£)	O'heads /profit (£)	Total (£)
Black enamel, steel						
3/4" - 20mm	nr	0.03	0.29	0.52	0.12	0.93
1" - 25mm	nr	0.03	0.29	0.65	0.14	1.08
20mm - 3/4"	nr	0.03	0.29	0.52	0.12	0.81
25mm - 1"	nr	0.03	0.29	0.65	0.14	1.08
Cable glands, brass fitted for TRS cables (up to 8.75mm diameter)						
20mm	nr	0.10	0.96	0.71	0.25	1.92
25mm	nr	0.13	1.24	1.33	0.39	2.96
32mm	nr	0.15	1.44	6.22	1.15	8.81
Reducers, fitted to conduit						
Male to female black enamel						
20 - 16mm	nr	0.02	0.19	0.65	0.13	0.97
25 - 16mm	nr	0.02	0.19	1.44	0.24	1.87
25 - 20mm	nr	0.02	0.19	0.77	0.14	1.10
32 - 20mm	nr	0.02	0.19	0.93	0.17	1.29
32 - 25mm	nr	0.02	0.19	0.93	0.17	1.29
1.5" - 20mm	nr	0.02	0.19	1.39	0.24	1.82
1.5" - 25mm	nr	0.02	0.19	1.39	0.24	1.82
1.5" - 32mm	nr	0.02	0.19	1.39	0.24	1.82
Male to female, galvanized						
20 - 16mm	nr	0.03	0.29	0.80	0.16	1.25
25 - 16mm	nr	0.03	0.29	1.78	0.31	2.38
25 - 20mm	nr	0.03	0.29	1.02	0.20	1.51
32 - 20mm	nr	0.03	0.29	1.18	0.22	1.69
32 - 25mm	nr	0.03	0.29	1.18	0.22	1.69
1.5" - 20mm	nr	0.03	0.29	1.77	0.31	2.37
1.5" - 25mm	nr	0.03	0.29	1.77	0.31	2.37
1.5" - 32mm	nr	0.03	0.29	1.77	0.31	2.37

M & E MEASUREMENT SERVICES

	Unit	Labour hours	Net labour (£)	Net material (£)	O'heads /profit (£)	Total (£)
Nipples, galvanized						
Screwed						
20mm	nr	0.03	0.29	0.31	0.09	0.69
25mm	nr	0.03	0.29	0.40	0.10	0.79
32mm	nr	0.03	0.29	0.78	0.16	1.23
Black enamel, screwed						
20mm	nr	0.03	0.29	0.24	0.08	0.61
25mm	nr	0.03	0.29	0.29	0.09	0.67
32mm	nr	0.03	0.29	0.55	0.13	0.97
Bonding						
20mm	nr	0.03	0.29	0.54	0.12	0.95
Unions, galvanized						
20mm	nr	0.08	0.77	3.21	0.60	4.58
25mm	nr	0.08	0.77	4.23	0.75	5.75
Black enamel						
20mm	nr	0.08	0.77	2.80	0.54	4.11
25mm	nr	0.08	0.77	3.50	0.64	4.91
Plugs						
Brass, slotted head						
20mm	nr	0.03	0.29	0.36	0.10	0.75
25mm	nr	0.03	0.29	0.55	0.13	0.97
32mm	nr	0.03	0.29	1.67	0.29	2.25
Iron, hexagonal head, galvanized						
20mm	nr	0.03	0.29	0.48	0.12	0.89
25mm	nr	0.03	0.29	0.69	0.15	1.13
32mm	nr	0.04	0.38	1.03	0.21	1.62

CONDUIT AND CABLE TRUNKING

Conduit fittings (cont'd)	Unit	Labour hours	Net labour (£)	Net material (£)	O'heads /profit (£)	Total (£)
Black enamel						
20mm	nr	0.03	0.29	0.42	0.11	0.82
25mm	nr	0.03	0.29	0.55	0.13	0.97
32mm	nr	0.04	0.38	0.78	0.17	1.33
Extension rings						
Malleable iron, galvanized						
6mm	nr	0.05	0.48	1.78	0.34	2.60
12mm	nr	0.05	0.48	1.47	0.29	2.24
19mm	nr	0.05	0.48	1.91	0.36	2.75
25mm	nr	0.05	0.48	2.19	0.40	3.07
32mm	nr	0.05	0.48	2.52	0.45	3.45
38mm	nr	0.05	0.48	2.79	0.49	3.76
Malleable iron, black enamel						
6mm	nr	0.05	0.48	1.44	0.29	2.21
12mm	nr	0.05	0.48	1.19	0.25	1.92
19mm	nr	0.05	0.48	1.55	0.30	2.33
25mm	nr	0.05	0.48	1.78	0.34	2.60
32mm	nr	0.05	0.48	1.99	0.37	2.84
38mm	nr	0.05	0.48	2.20	0.40	3.08
Clips						
Steel, galvanized						
20mm	nr	0.12	1.15	0.05	0.18	1.38
25mm	nr	0.12	1.15	0.06	0.18	1.39
Black enamel						
20mm	nr	0.12	1.15	0.04	0.18	1.37
25mm	nr	0.12	1.15	0.05	0.18	1.38
Hooks, male, screwed to backgrounds, galvanized						
20mm	nr	0.08	0.77	1.32	0.31	2.40
25mm	nr	0.08	0.77	1.72	0.37	2.86

	Unit	Labour hours	Net labour (£)	Net material (£)	O'heads /profit (£)	Total (£)
Black enamel						
20mm	nr	0.08	0.77	1.03	0.27	2.07
25mm	nr	0.08	0.77	1.36	0.32	2.45
Covers, fixed to backgrounds with screws						
Small circular, galvanized						
pendant plate, 20mm	nr	0.03	0.29	1.71	0.30	2.30
pendant plate, 25mm	nr	0.03	0.29	2.21	0.38	2.88
hook plate	nr	0.03	0.29	1.18	0.22	1.69
ball and socket, 20mm	nr	0.03	0.29	4.70	0.75	5.74
light steel	nr	0.03	0.29	0.19	0.07	0.55
heavy steel	nr	0.03	0.29	0.37	0.10	0.76
overlapping	nr	0.03	0.29	0.35	0.10	0.64
Small circular, black enamel						
pendant plate, 20mm	nr	0.03	0.29	1.35	0.25	1.89
pendant plate, 25mm	nr	0.03	0.29	1.75	0.31	2.35
hook plate	nr	0.03	0.29	0.91	0.18	1.38
ball and socket plate	nr	0.03	0.29	3.78	0.61	4.68
light steel	nr	0.03	0.29	0.15	0.07	0.51
heavy steel	nr	0.03	0.29	0.25	0.08	0.62
overlapping	nr	0.03	0.29	0.28	0.09	0.57
Large circular, galvanized						
flat	nr	0.03	0.29	0.75	0.16	1.20
pendant, 20mm	nr	0.03	0.29	4.11	0.66	5.06
Large circular, black enamel						
flat	nr	0.03	0.29	0.56	0.13	0.98
pendant, 20mm	nr	0.03	0.29	3.40	0.55	4.24
Washers, fitted						
Brass compression						
20mm	nr	0.03	0.29	0.10	0.06	0.45
25mm	nr	0.03	0.29	0.12	0.06	0.47

CONDUIT AND CABLE TRUNKING

	Unit	Labour hours	Net labour (£)	Net material (£)	O'heads /profit (£)	Total (£)
PVC conduit fittings						

A discount of 40% has been incorporated within the material costs for PVC fittings

PVC fittings for connection to PVC conduit including any cutting and threading

Bends

Normal, heavy gauge black or white

	Unit	Labour hours	Net labour (£)	Net material (£)	O'heads /profit (£)	Total (£)
20mm	nr	0.07	0.67	1.23	0.28	2.18
25mm	nr	0.07	0.67	1.66	0.35	2.68
32mm	nr	0.08	0.77	2.60	0.51	3.88

Normal, light gauge, white

16mm	nr	0.07	0.67	0.42	0.16	1.25
20mm	nr	0.07	0.67	0.46	0.17	1.30
25mm	nr	0.08	0.77	0.77	0.23	1.77
32mm	nr	0.08	0.77	1.43	0.33	2.53

Inspection, black or white

20mm	nr	0.10	0.96	1.05	0.30	2.31
25mm	nr	0.12	1.15	1.99	0.47	3.61

Tees

Inspection, black or white

20mm	nr	0.15	1.44	1.05	0.37	2.86
25mm	nr	0.17	1.63	2.12	0.56	4.31

Boxes adaptable, moulded, black or white

75 x 75 x 41mm	nr	0.20	1.91	1.43	0.50	3.34
75 x 75 x 53mm	nr	0.20	1.91	1.89	0.57	3.80
100 x 100 x 75mm	nr	0.20	1.91	3.18	0.76	5.09
100 x 100 x 50mm	nr	0.20	1.91	3.02	0.74	4.93

	Unit	Labour hours	Net labour (£)	Net material (£)	O'heads /profit (£)	Total (£)
100 x 75 x 50mm	nr	0.20	1.91	3.10	0.75	5.01
150 x 100 x 50mm	nr	0.20	1.91	3.77	0.85	5.68
150 x 150 x 75mm	nr	0.20	1.91	4.08	0.90	5.99
225 x 225 x 75mm	nr	0.33	3.16	6.95	1.52	10.11

Circular junction, black or white

Back outlet

20mm	nr	0.09	0.86	1.20	0.31	2.37
25mm	nr	0.09	0.86	1.93	0.42	3.21

Terminal - 1 way

20mm	nr	0.09	0.86	0.89	0.26	2.01
25mm	nr	0.09	0.86	1.40	0.34	2.60

Through - 2 way

20mm	nr	0.09	0.86	1.01	0.28	2.15
25mm	nr	0.09	0.86	1.52	0.36	2.74

Angle - 2 way

20mm	nr	0.09	0.86	1.01	0.28	2.15
25mm	nr	0.09	0.86	1.52	0.36	2.74

Tee - 3 way

20mm	nr	0.09	0.86	1.08	0.29	2.23
25mm	nr	0.09	0.86	1.68	0.38	2.92

Intersection - 4 way

20mm	nr	0.09	0.86	1.27	0.32	2.45
25mm	nr	0.09	0.86	1.91	0.42	3.19

Circular junction, black or white

One way and back outlet

20mm	nr	0.09	0.86	1.47	0.35	2.68
25mm	nr	0.09	0.86	2.16	0.45	3.47

CONDUIT AND CABLE TRUNKING

Conduit fittings (cont'd)	Unit	Labour hours	Net labour (£)	Net material (£)	O'heads /profit (£)	Total (£)
Two way and back outlet						
20mm	nr	0.09	0.86	1.63	0.37	2.86
25mm	nr	0.09	0.86	2.19	0.46	3.51
Angle and back outlet						
20mm	nr	0.09	0.86	1.63	0.37	2.86
25mm	nr	0.09	0.86	2.19	0.46	3.51
Three way and back outlet						
20mm	nr	0.09	0.86	1.76	0.39	3.01
25mm	nr	0.09	0.86	2.31	0.48	3.65
Four way and back outlet						
20mm	nr	0.09	0.86	1.77	0.39	3.02
25mm	nr	0.09	0.86	2.48	0.50	3.84
Tangent angle						
20mm	nr	0.09	0.86	1.29	0.32	2.47
25mm	nr	0.09	0.86	1.89	0.41	3.16
Tangent - 3 way or 'T'						
20mm	nr	0.09	0.86	1.50	0.35	2.71
25mm	nr	0.09	0.86	2.03	0.43	3.32
Branch - 2 way or 'U'						
20mm	nr	0.09	0.86	1.89	0.41	3.16
25mm	nr	0.09	0.86	2.42	0.49	3.77
Branch - 3 way or 'Y'						
20mm	nr	0.09	0.86	2.32	0.48	3.66
25mm	nr	0.09	0.86	2.88	0.56	4.30
Twin - through way or 'H'						
20mm	nr	0.09	0.86	2.43	0.49	3.78
25mm	nr	0.09	0.86	3.01	0.58	4.45

M & E MEASUREMENT SERVICES

	Unit	Labour hours	Net labour (£)	Net material (£)	O'heads /profit (£)	Total (£)
Circular junction with swivel lug, black or white						
Back outlet (fixed external lug)						
20mm	nr	0.09	0.86	1.39	0.34	2.59
25mm	nr	0.09	0.86	1.50	0.35	2.71
Terminal - 1 way (fixed external lug)						
20mm	nr	0.09	0.86	2.27	0.47	3.60
25mm	nr	0.09	0.86	2.15	0.45	3.46
Through - 2 way						
20mm	nr	0.09	0.86	1.73	0.39	2.98
25mm	nr	0.09	0.86	2.28	0.47	3.61
Angle - 2 way						
20mm	nr	0.09	0.86	1.74	0.39	2.99
25mm	nr	0.09	0.86	2.28	0.47	3.61
Tee - 3 way						
20mm	nr	0.09	0.86	1.83	0.40	3.09
25mm	nr	0.09	0.86	2.43	0.49	3.78
Intersection - 4 way						
20mm	nr	0.09	0.86	2.01	0.43	3.30
25mm	nr	0.09	0.86	2.65	0.53	4.04
One way and back outlet						
20mm	nr	0.09	0.86	2.23	0.46	3.55
25mm	nr	0.09	0.86	2.84	0.55	4.25
Two way and back outlet						
20mm	nr	0.09	0.86	2.37	0.48	3.71
25mm	nr	0.09	0.86	2.85	0.56	4.27

CONDUIT AND CABLE TRUNKING

Conduit fittings (cont'd)	Unit	Labour hours	Net labour (£)	Net material (£)	O'heads /profit (£)	Total (£)
Angle and back outlet						
20mm	nr	0.09	0.86	2.37	0.48	3.71
25mm	nr	0.09	0.86	2.85	0.56	4.27
Three way and back outlet						
20mm	nr	0.09	0.86	2.58	0.52	3.96
25mm	nr	0.09	0.86	2.98	0.58	4.42
Four way and back outlet						
20mm	nr	0.09	0.86	2.46	0.50	3.82
25mm	nr	0.09	0.86	3.13	0.60	4.59
Tangent angle						
20mm	nr	0.09	0.86	2.11	0.45	3.42
25mm	nr	0.09	0.86	2.77	0.54	4.17
Tangent - 3 way or 'T'						
20mm	nr	0.09	0.86	2.27	0.47	3.60
25mm	nr	0.09	0.86	2.89	0.56	4.31
Branch - 2 way or 'U'						
20mm	nr	0.07	0.67	2.42	0.46	3.55
25mm	nr	0.08	0.77	2.95	0.56	4.28
Branch - 3 way or 'Y'						
20mm	nr	0.07	0.67	2.49	0.47	3.63
25mm	nr	0.08	0.77	3.04	0.57	4.38
Twin - through way or 'H'						
20mm	nr	0.07	0.67	2.51	0.48	3.66
25mm	nr	0.07	0.67	2.99	0.55	4.21

	Unit	Labour hours	Net labour (£)	Net material (£)	O'heads /profit (£)	Total (£)
Square junction boxes connected to conduit and fixed to background requiring drilling, plugging and screwing						
Through						
75 x 75 x 32mm	nr	0.25	2.39	4.49	1.03	7.91
75 x 75 x 38mm	nr	0.25	2.39	7.40	1.47	11.26
100 x 100 x 50mm	nr	0.25	2.39	14.08	2.47	18.94
Angle						
75 x 75 x 32mm	nr	0.20	1.91	5.11	1.05	8.07
75 x 75 x 38mm	nr	0.20	1.91	7.39	1.39	10.69
100 x 100 x 50mm	nr	0.20	1.91	14.08	2.40	18.39
Tee						
75 x 75 x 32mm	nr	0.20	1.91	6.10	1.20	9.21
75 x 75 x 38mm	nr	0.20	1.91	9.11	1.65	12.67
100 x 100 x 50mm	nr	0.20	1.91	17.60	2.93	22.44
uPVC outlets fixed to backgrounds requiring drilling, plugging and screwing and connections to conduit						
Flush socket outlet, heavy gauge, fixed lugs						
1 gang						
35mm deep	nr	0.15	1.44	1.03	0.37	2.84
47mm deep	nr	0.15	1.44	1.24	0.40	3.08
2 gang						
30mm deep	nr	0.15	1.44	2.07	0.53	4.04
44mm deep	nr	0.15	1.44	2.26	0.55	4.25

Conduit fittings (cont'd)	Unit	Labour hours	Net labour (£)	Net material (£)	O'heads /profit (£)	Total (£)
Flush socket outlet, standard fixed lugs, 35mm deep						
1 gang	nr	0.15	1.44	1.22	0.40	3.06
2 gang	nr	0.15	1.44	1.81	0.49	3.74
Surface switch outlet						
25mm deep						
1 gang	nr	0.15	1.44	0.99	0.36	2.79
2 gang	nr	0.15	1.44	2.00	0.52	3.96
32mm deep						
1 gang	nr	0.15	1.44	1.20	0.40	3.04
2 gang	nr	0.15	1.44	2.21	0.55	4.20
PVC saddles, fixed to backgrounds requiring drilling, plugging and screwing						
Strap, black or white						
16mm	nr	0.12	1.15	0.14	0.19	1.48
20mm	nr	0.12	1.15	0.14	0.19	1.48
25mm	nr	0.12	1.15	0.18	0.20	1.53
32mm	nr	0.15	1.44	0.22	0.25	1.91
Spacer bar, black or white						
16mm	nr	0.13	1.24	0.32	0.23	1.79
20mm	nr	0.13	1.24	0.31	0.23	1.78
25mm	nr	0.13	1.24	0.40	0.25	1.89
32mm	nr	0.17	1.63	0.83	0.37	2.83
Spring clip, black or white						
16mm	nr	0.12	1.15	0.23	0.21	1.59
20mm	nr	0.12	1.15	0.26	0.21	1.62
25mm	nr	0.12	1.15	0.44	0.24	1.83
32mm	nr	0.15	1.44	0.73	0.33	2.50

M & E MEASUREMENT SERVICES

	Unit	Labour hours	Net labour (£)	Net material (£)	O'heads /profit (£)	Total (£)
Covers fixed to backgrounds with screws						
circular protector	nr	0.03	0.29	0.29	0.09	0.67
pendant dome, conduit thread, 20mm	nr	0.03	0.29	1.10	0.21	1.39
square box, 75mm x 75mm	nr	0.03	0.29	0.33	0.09	0.62
Couplings fitted to conduit						
Expansion, black or white						
16mm	nr	0.02	0.19	0.50	0.10	0.79
20mm	nr	0.02	0.19	0.53	0.11	0.83
25mm	nr	0.02	0.19	0.63	0.12	0.94
32mm	nr	0.03	0.29	1.50	0.27	2.06
Inspection, black or white						
20mm	nr	0.02	0.19	2.38	0.39	2.96
25mm	nr	0.02	0.19	2.74	0.44	3.37
Heavy gauge						
16mm	nr	0.02	0.19	0.27	0.07	0.53
20mm	nr	0.02	0.19	0.25	0.07	0.51
25mm	nr	0.02	0.19	0.33	0.08	0.60
32mm	nr	0.03	0.29	0.68	0.15	1.12
Bushes						
Bellmouth						
20mm	nr	0.02	0.19	0.12	0.05	0.36
25mm	nr	0.02	0.19	0.18	0.06	0.43
Sleeves for above						
20mm	nr	0.02	0.19	0.19	0.06	0.44
25mm	nr	0.02	0.19	0.31	0.07	0.57

145

CONDUIT AND CABLE TRUNKING

Conduit fittings (cont'd)	Unit	Labour hours	Net labour (£)	Net material (£)	O'heads /profit (£)	Total (£)
Female, screwed						
20mm	nr	0.02	0.19	0.31	0.07	0.57
25mm	nr	0.02	0.19	0.42	0.09	0.70
32mm	nr	0.03	0.29	0.71	0.15	1.15
Male, plain						
20mm	nr	0.02	0.19	0.74	0.14	1.07
25mm	nr	0.02	0.19	0.81	0.15	1.15
Male, screwed						
16mm	nr	0.02	0.19	0.11	0.04	0.34
20mm	nr	0.02	0.19	0.12	0.05	0.36
25mm	nr	0.02	0.19	0.30	0.07	0.56
32mm	nr	0.03	0.29	0.46	0.11	0.86
Glands, screwed						
Compression						
16mm	nr	0.30	2.87	1.24	0.62	4.73
20mm	nr	0.30	2.87	0.84	0.56	4.27
25mm	nr	0.30	2.87	2.13	0.75	5.75
32mm	nr	0.30	2.87	5.07	1.19	9.13
Locknuts						
16mm	nr	0.02	0.19	0.24	0.06	0.43
20mm	nr	0.02	0.19	0.24	0.06	0.49
25mm	nr	0.02	0.19	0.34	0.08	0.61
32mm	nr	0.02	0.19	0.54	0.11	0.84
Reducers						
Plain, black or white						
20 x 16mm	nr	0.03	0.29	0.59	0.13	1.01
25 x 20mm	nr	0.03	0.29	0.59	0.13	1.01
32 x 25mm	nr	0.03	0.29	0.78	0.16	1.23

M & E MEASUREMENT SERVICES

	Unit	Labour hours	Net labour (£)	Net material (£)	O'heads /profit (£)	Total (£)
Lockrings						
20mm	nr	0.02	0.19	0.24	0.06	0.49
25mm	nr	0.02	0.19	0.26	0.07	0.52
32mm	nr	0.02	0.19	0.35	0.08	0.62
Extension rings						
With brass inserts						
6mm	nr	0.05	0.48	0.50	0.15	1.13
12.5mm	nr	0.05	0.48	0.64	0.17	1.29
19mm	nr	0.05	0.48	0.83	0.20	1.51
25mm	nr	0.05	0.48	0.98	0.22	1.68
32mm	nr	0.05	0.48	1.04	0.23	1.75
Clips						
Circular conduit						
16mm	nr	0.12	1.15	0.36	0.23	1.74
20mm	nr	0.12	1.15	0.38	0.23	1.76
25mm	nr	0.12	1.15	0.44	0.24	1.83
32mm	nr	0.15	1.44	0.71	0.32	2.47
Terminals						
Brass earthing						
5 amp	nr	0.08	0.77	0.42	0.18	1.37
15 amp	nr	0.08	0.77	0.52	0.19	1.48
Adaptors						
Clip-in						
20mm	nr	0.03	0.29	0.35	0.10	0.74
25mm	nr	0.03	0.29	0.48	0.12	0.89
Male thread with locking rings						
16mm	nr	0.03	0.29	0.36	0.10	0.75
20mm	nr	0.03	0.29	0.29	0.09	0.67
25mm	nr	0.03	0.29	0.54	0.12	0.95
32mm	nr	0.03	0.29	0.97	0.19	1.45

CONDUIT AND CABLE TRUNKING

Conduit fittings (cont'd)	Unit	Labour hours	Net labour (£)	Net material (£)	O'heads /profit (£)	Total (£)
Sundries						
Adhesive - 1 litre	nr	0.00	0.00	19.59	2.94	22.53
Bending springs						
Light gauge						
16mm	nr	0.00	0.00	10.92	1.64	12.56
20mm	nr	0.00	0.00	10.92	1.64	12.56
25mm	nr	0.00	0.00	16.10	2.42	18.52
32mm	nr	0.00	0.00	31.44	4.72	36.16
Heavy gauge						
16mm	nr	0.00	0.00	10.92	1.64	12.56
20mm	nr	0.00	0.00	10.92	1.64	12.56
25mm	nr	0.00	0.00	16.10	2.42	18.52
32mm	nr	0.00	0.00	31.44	4.72	36.16

	Unit	Labour hours	Net labour (£)	Net material (£)	O'heads /profit (£)	Total (£)

STEEL CABLE TRAY

General notes

1. All labour times include for installing surface fixed cable tray using site fabrication methods, i.e. cutting, drilling and connecting

2. All labour times include for drilling and bolting fittings to tray

3. All labour times include for all fixings to brackets for ladder rack

4. A discount of 40% has been incorporated within the net material costs in this section

5. 1.5% waste factor has been added to the material costs of straight tray

6. All labour times and material costs include standard coupling joints, fixings and earth continuity straps measured in the running length. Supports measured elsewhere

Hot dip galvanized light/medium duty cable tray

Straight tray

Width mm	Gauge mm	Unit	Labour hours	Net labour (£)	Net material (£)	O'heads /profit (£)	Total (£)
50	1.0	m	0.20	1.91	1.65	0.53	3.56
50	1.5	m	0.20	1.91	2.16	0.61	4.07
75	1.0	m	0.25	2.39	1.94	0.65	4.33
75	1.5	m	0.25	2.39	2.28	0.70	4.67
100	1.0	m	0.28	2.68	2.18	0.73	4.86

Steel cable tray (cont'd)		Unit	Labour hours	Net labour (£)	Net material (£)	O'heads /profit (£)	Total (£)
100	1.5	m	0.28	2.68	2.75	0.81	5.43
150	1.0	m	0.33	3.16	2.79	0.89	5.95
150	1.5	m	0.33	3.16	3.87	1.05	7.03
225	1.2	m	0.40	3.83	4.33	1.22	8.16
225	1.5	m	0.40	3.83	5.14	1.35	8.97
300	1.5	m	0.57	5.45	6.75	1.83	12.20
450	1.5	m	0.57	5.45	10.71	2.42	16.16
450	2.0	m	0.57	5.45	11.86	2.60	17.31
600	2.0	m	0.80	7.66	14.74	3.36	22.40
750	2.0	m	0.87	8.33	19.64	4.20	27.97
900	2.0	m	0.98	9.38	24.51	5.08	33.89

Flat bends

Width mm	Gauge mm						
50	1.5	nr	0.20	1.91	2.45	0.65	4.36
75	1.5	nr	0.20	1.91	2.57	0.67	4.48
100	1.5	nr	0.20	1.91	3.17	0.76	5.08
150	1.5	nr	0.25	2.39	3.29	0.85	5.68
225	1.5	nr	0.30	2.87	4.79	1.15	7.66
300	1.5	nr	0.42	4.02	6.77	1.62	10.79
450	2.0	nr	0.57	5.45	11.39	2.53	16.84
600	2.0	nr	0.75	7.18	18.05	3.78	25.23
750	2.0	nr	0.83	7.94	25.91	5.08	33.85
900	2.0	nr	0.97	9.28	38.39	7.15	47.67

Outside risers

Width mm	Gauge mm						
50	1.5	nr	0.20	1.91	4.91	1.02	6.82
75	1.5	nr	0.20	1.91	5.57	1.12	7.48
100	1.5	nr	0.20	1.91	6.47	1.26	8.38
150	1.5	nr	0.25	2.39	8.69	1.66	11.08
225	1.5	nr	0.30	2.87	11.15	2.10	14.02
300	1.5	nr	0.42	4.02	14.93	2.84	18.95
450	2.0	nr	0.57	5.45	20.69	3.92	26.14
600	2.0	nr	0.75	7.18	26.57	5.06	33.75
750	2.0	nr	0.83	7.94	35.09	6.45	43.03
900	2.0	nr	0.97	9.28	41.63	7.64	50.91

M & E MEASUREMENT SERVICES

		Unit	Labour hours	Net labour (£)	Net material (£)	O'heads /profit (£)	Total (£)
Inside risers							
Width	Gauge						
mm	mm						
50	1.5	nr	0.20	1.91	4.91	1.02	6.82
75	1.5	nr	0.20	1.91	5.57	1.12	7.48
100	1.5	nr	0.20	1.91	6.47	1.26	8.38
150	1.5	nr	0.25	2.39	8.69	1.66	11.08
225	1.5	nr	0.30	2.87	11.15	2.10	14.02
300	1.5	nr	0.42	4.02	14.93	2.84	18.95
450	2.0	nr	0.57	5.45	20.69	3.92	26.14
600	2.0	nr	0.75	7.18	26.57	5.06	33.75
750	2.0	nr	0.83	7.94	35.09	6.45	43.03
900	2.0	nr	0.97	9.28	41.63	7.64	50.91
Tees							
Width	Gauge						
mm	mm						
50	1.5	nr	0.30	2.87	3.41	0.94	7.22
75	1.5	nr	0.30	2.87	3.59	0.97	7.43
100	1.5	nr	0.30	2.87	4.01	1.03	7.91
150	1.5	nr	0.33	3.16	4.61	1.17	8.94
225	1.5	nr	0.42	4.02	6.77	1.62	12.41
300	1.5	nr	0.62	5.93	9.95	2.38	18.26
450	2.0	nr	0.85	8.13	16.89	3.75	28.77
600	2.0	nr	1.12	10.72	25.91	5.49	42.12
750	2.0	nr	1.17	11.20	39.71	7.64	58.55
900	2.0	nr	1.47	14.07	56.75	10.62	81.44
4 way cross piece							
Width	Gauge						
mm	mm						
50	1.5	nr	0.30	2.87	5.09	1.19	9.15
75	1.5	nr	0.30	2.87	5.27	1.22	9.36
100	1.5	nr	0.30	2.87	5.93	1.32	10.12
150	1.5	nr	0.33	3.16	6.89	1.51	11.56
225	1.5	nr	0.42	4.02	9.83	2.08	15.93
300	1.5	nr	0.62	5.93	14.75	3.10	23.78
450	2.0	nr	0.85	8.13	25.25	5.01	38.39
600	2.0	nr	1.12	10.72	34.43	6.77	51.92
750	2.0	nr	1.17	11.20	52.79	9.60	73.59
900	2.0	nr	1.47	14.07	77.75	13.77	105.59

CONDUIT AND CABLE TRUNKING

Steel cable tray (cont'd)	Unit	Labour hours	Net labour (£)	Net material (£)	O'heads /profit (£)	Total (£)
Reducers						
Width Gauge						
mm mm						
75 1.5	nr	0.17	1.63	4.49	0.92	7.04
100 1.5	nr	0.17	1.63	5.03	1.00	7.66
150 1.5	nr	0.17	1.63	6.89	1.28	9.80
225 1.5	nr	0.23	2.20	9.41	1.74	13.35
300 1.5	nr	0.23	2.20	12.35	2.18	16.73
450 2.0	nr	0.37	3.54	19.01	3.38	25.93
600 2.0	nr	0.37	3.54	24.29	4.17	32.00
750 2.0	nr	0.58	5.55	31.19	5.51	42.25
900 2.0	nr	0.58	5.55	36.41	6.29	48.25
Epoxy coated light/medium duty cable tray						
Straight tray						
Width Gauge						
mm mm						
50 1.5	m	0.20	1.91	4.03	0.89	6.83
75 1.5	m	0.25	2.39	4.17	0.98	7.54
100 1.5	m	0.28	2.68	4.80	1.12	8.60
150 1.5	m	0.33	3.16	6.65	1.47	11.28
225 1.5	m	0.40	3.83	8.78	1.89	14.50
300 1.5	m	0.50	4.79	11.26	2.41	18.46
450 2.0	m	0.57	5.45	14.98	3.06	23.49
610 2.0	m	0.80	7.66	22.65	4.55	34.86
762 2.0	m	0.87	8.33	25.96	5.14	39.43
915 2.0	m	0.98	9.38	30.68	6.01	46.07
Flat bends						
Width Gauge						
mm mm						
50 1.5	nr	0.20	1.91	3.41	0.80	6.12
75 1.5	nr	0.20	1.91	3.71	0.84	6.46
100 1.5	nr	0.20	1.91	4.19	0.92	7.02
150 1.5	nr	0.25	2.39	5.57	1.19	9.15
225 1.5	nr	0.30	2.87	8.15	1.65	12.67
300 1.5	nr	0.42	4.02	11.39	2.31	17.72
400 2.0	nr	0.57	5.45	19.85	3.79	29.09
600 2.0	nr	0.75	7.18	27.83	5.25	40.26
750 2.0	nr	0.83	7.94	36.29	6.63	50.86
900 2.0	nr	0.97	9.28	46.79	8.41	64.48

Steel cable tray (cont'd)	Unit	Labour hours	Net labour (£)	Net material (£)	O'heads /profit (£)	Total (£)
Outside risers						
Width Gauge						
mm mm						
50 1.5	nr	0.20	1.91	7.43	1.40	10.74
75 1.5	nr	0.20	1.91	7.61	1.43	10.95
100 1.5	nr	0.20	1.91	8.81	1.61	12.33
150 1.5	nr	0.25	2.39	13.01	2.31	17.71
225 1.5	nr	0.30	2.87	17.03	2.98	22.88
300 1.5	nr	0.42	4.02	21.41	3.81	29.24
450 2.0	nr	0.57	5.45	26.51	4.79	36.75
600 2.0	nr	0.75	7.18	39.17	6.95	53.30
750 2.0	nr	0.83	7.94	44.69	7.89	60.52
900 2.0	nr	0.97	9.28	52.67	9.29	71.24
Inside risers						
Width Gauge						
mm mm						
50 1.5	nr	0.20	1.91	7.43	1.40	10.74
75 1.5	nr	0.20	1.91	7.61	1.43	10.95
100 1.5	nr	0.20	1.91	8.81	1.61	12.33
150 1.5	nr	0.25	2.39	13.01	2.31	17.71
225 1.5	nr	0.30	2.87	17.03	2.98	22.88
300 1.5	nr	0.42	4.02	21.41	3.81	29.24
450 2.0	nr	0.57	5.45	26.51	4.79	36.75
600 2.0	nr	0.75	7.18	39.17	6.95	53.30
750 2.0	nr	0.83	7.94	44.69	7.89	60.52
900 2.0	nr	0.97	9.28	52.67	9.29	71.24
Tees						
Width Gauge						
mm mm						
50 1.5	nr	0.30	2.87	5.03	1.19	9.09
75 1.5	nr	0.30	2.87	5.51	1.26	9.64
110 1.5	nr	0.30	2.87	6.17	1.36	10.40
150 1.5	nr	0.33	3.16	8.21	1.71	13.08
225 1.5	nr	0.42	4.02	12.17	2.43	18.62
300 1.5	nr	0.62	5.93	16.85	3.42	26.20
450 2.0	nr	0.85	8.13	28.43	5.48	42.04
600 2.0	nr	1.12	10.72	41.57	7.84	60.13
750 2.0	nr	1.17	11.20	54.17	9.81	75.18
900 2.0	nr	1.47	14.07	70.01	12.61	96.69

CONDUIT AND CABLE TRUNKING

		Unit	Labour hours	Net labour (£)	Net material (£)	O'heads /profit (£)	Total (£)
4 way cross piece							
Width	Gauge						
mm	mm						
50	1.5	nr	0.30	2.87	7.43	1.54	11.84
75	1.5	nr	0.30	2.87	8.03	1.63	12.53
100	1.5	nr	0.30	2.87	9.05	1.79	13.71
150	1.5	nr	0.33	3.16	12.23	2.31	17.70
225	1.5	nr	0.42	4.02	17.87	3.28	25.17
300	1.5	nr	0.62	5.93	25.19	4.67	35.79
450	2.0	nr	0.85	8.13	42.89	7.65	58.67
610	2.0	nr	1.12	10.72	62.27	10.95	83.94
750	2.0	nr	1.17	11.20	81.29	13.87	106.36
900	2.0	nr	1.47	14.07	104.87	17.84	136.78
Reducer							
Width	Gauge						
mm	mm						
75	1.5	nr	0.17	1.63	7.61	1.39	10.63
100	1.5	nr	0.17	1.63	8.81	1.57	12.01
150	1.5	nr	0.17	1.63	12.23	2.08	15.94
225	1.5	nr	0.23	2.20	14.93	2.57	19.70
300	1.5	nr	0.23	2.20	20.63	3.42	26.25
450	2.0	nr	0.37	3.54	27.29	4.62	35.45
600	2.0	nr	0.37	3.54	33.29	5.52	42.35
750	2.0	nr	0.58	5.55	36.93	6.37	48.85
900	2.0	nr	0.58	5.55	39.83	6.81	52.19
Pre-galvanized light/medium duty cable tray							
Straight tray							
Width	Gauge						
mm	mm						
50	1.5	m	0.20	1.91	1.07	0.45	3.43
75	1.5	m	0.25	2.39	1.33	0.56	4.28
100	1.5	m	0.28	2.68	1.64	0.65	4.97
150	1.5	m	0.33	3.16	2.19	0.80	6.15
225	1.5	m	0.40	3.83	4.03	1.18	9.04
300	2.0	m	0.50	4.79	5.57	1.55	11.91
450	1.5	m	0.57	5.45	8.46	2.09	16.00
600	2.0	m	0.80	7.66	14.18	3.28	25.12
750	2.0	m	0.87	8.33	17.53	3.88	29.74
900	2.0	m	0.98	9.38	20.86	4.54	34.78

M & E MEASUREMENT SERVICES

		Unit	Labour hours	Net labour (£)	Net material (£)	O'heads /profit (£)	Total (£)
Flat bends							
Width	Gauge						
mm	mm						
50	1.5	nr	0.20	1.91	2.03	0.59	4.53
75	1.5	nr	0.20	1.91	2.15	0.61	4.67
100	1.5	nr	0.20	1.91	2.27	0.63	4.81
150	1.5	nr	0.25	2.39	2.75	0.92	5.14
225	1.5	nr	0.30	2.87	3.71	0.99	7.57
300	1.5	nr	0.42	4.02	5.69	1.46	11.17
450	2.0	nr	0.57	5.45	10.07	2.33	17.85
600	2.0	nr	0.75	7.18	15.71	3.43	26.32
750	2.0	nr	0.83	7.94	22.49	4.56	34.99
900	2.0	nr	0.97	9.28	28.73	5.70	43.71
Outside risers							
Width	Gauge						
mm	mm						
50	1.5	nr	0.20	1.91	3.41	0.80	6.12
75	1.5	nr	0.20	1.91	3.71	0.84	6.46
100	1.5	nr	0.20	1.91	4.31	0.93	7.15
150	1.5	nr	0.25	2.39	5.69	1.21	9.29
225	1.5	nr	0.30	2.87	7.67	1.58	12.12
300	1.5	nr	0.42	4.02	8.27	1.84	14.13
450	2.0	nr	0.57	5.45	14.87	3.05	23.37
600	2.0	nr	0.75	7.18	18.89	3.91	29.98
750	2.0	nr	0.83	7.94	24.83	4.92	37.69
900	2.0	nr	0.97	9.28	29.87	5.87	45.02
Inside risers							
Width	Gauge						
mm	mm						
50	1.5	nr	0.30	2.87	3.41	0.94	7.22
75	1.5	nr	0.30	2.87	3.71	0.99	7.57
100	1.5	nr	0.30	2.87	4.31	1.08	8.26
150	1.5	nr	0.33	3.16	5.69	1.33	10.18
225	1.5	nr	0.45	4.31	7.67	1.80	13.78
300	1.5	nr	0.62	5.93	8.27	2.13	16.33
450	2.0	nr	0.85	8.13	14.87	3.45	26.45
600	2.0	nr	1.12	10.72	18.89	4.44	34.05
750	2.0	nr	1.17	11.20	24.83	5.40	41.43
900	2.0	nr	1.47	14.07	29.87	6.59	50.53

Steel cable tray (cont'd)		Unit	Labour hours	Net labour (£)	Net material (£)	O'heads /profit (£)	Total (£)
Tees							
Width	Gauge						
mm	mm						
50	1.5	nr	0.30	2.87	2.93	0.87	6.67
75	1.5	nr	0.30	2.87	3.05	0.89	6.81
100	1.5	nr	0.30	2.87	3.23	0.91	7.01
150	1.5	nr	0.33	3.16	4.13	1.09	8.38
225	1.5	nr	0.45	4.31	5.75	1.51	11.57
300	1.5	nr	0.62	5.93	8.39	2.15	16.47
450	2.0	nr	0.85	8.13	14.45	3.39	25.97
600	2.0	nr	1.12	10.72	21.23	4.79	36.74
750	2.0	nr	1.17	11.20	30.71	6.29	48.20
900	2.0	nr	1.47	14.07	44.93	8.85	67.85
4 way cross piece							
Width	Gauge						
mm	mm						
50	1.5	nr	0.30	2.87	4.37	1.09	8.33
75	1.5	nr	0.30	2.87	4.43	1.09	8.39
100	1.5	nr	0.30	2.87	4.73	1.14	8.74
150	1.5	nr	0.33	3.16	5.93	1.36	10.45
225	1.5	nr	0.45	4.31	8.33	1.90	14.54
300	1.5	nr	0.62	5.93	11.81	2.66	20.40
450	2.0	nr	0.85	8.13	21.71	4.48	34.32
600	2.0	nr	1.12	10.72	27.71	5.76	44.19
750	2.0	nr	1.17	11.20	39.89	7.66	58.75
900	2.0	nr	1.47	14.07	50.75	9.72	74.54
Reducers							
Width	Gauge						
mm	mm						
75	1.5	nr	0.17	1.63	3.77	0.81	6.21
100	1.5	nr	0.17	1.63	4.13	0.86	6.62
150	1.5	nr	0.23	2.20	5.51	1.16	8.87
225	1.5	nr	0.23	2.20	6.95	1.37	10.52
300	1.5	nr	0.37	3.54	9.53	1.96	15.03
450	2.0	nr	0.37	3.54	14.45	2.70	20.69
600	2.0	nr	0.58	5.55	18.41	3.59	27.55
750	2.0	nr	0.58	5.55	19.55	3.77	28.87
900	2.0	nr	0.80	7.66	20.87	4.28	32.81

	Unit	Labour hours	Net labour (£)	Net material (£)	O'heads /profit (£)	Total (£)
Hot dip galvanized heavy duty cable tray						
Straight tray						
Width Gauge						
mm mm						
100 1.5	m	0.28	2.68	3.98	1.00	7.66
150 1.5	m	0.33	3.16	4.56	1.16	8.88
225 1.5	m	0.40	3.83	5.80	1.44	11.07
300 1.5	m	0.50	4.79	7.81	1.89	14.49
450 2.0	m	0.57	5.45	13.24	2.80	21.49
600 2.0	m	0.80	7.66	15.73	3.51	26.90
Flat bends						
Width Gauge						
mm mm						
100 1.5	nr	0.20	1.91	3.65	0.83	6.39
150 1.5	nr	0.25	2.39	4.07	0.97	7.43
225 1.5	nr	0.30	2.87	5.51	1.26	9.64
300 1.5	nr	0.42	4.02	7.43	1.72	13.17
450 2.0	nr	0.57	5.45	13.97	2.91	22.33
600 2.0	nr	0.75	7.18	19.43	3.99	30.60
Outside risers						
Width Gauge						
mm mm						
100 1.5	nr	0.20	1.91	8.09	1.50	11.50
150 1.5	nr	0.25	2.39	9.05	1.72	13.16
225 1.5	nr	0.30	2.87	11.51	2.16	16.54
300 1.5	nr	0.42	4.02	13.55	2.64	20.21
450 2.0	nr	0.57	5.45	20.27	3.86	29.58
600 2.0	nr	0.75	7.18	28.73	5.39	41.30
Inside risers						
Width Gauge						
mm mm						
102 1.5	nr	0.20	1.91	8.09	1.50	11.50
150 1.5	nr	0.25	2.39	9.05	1.72	13.16
225 1.5	nr	0.30	2.87	11.51	2.16	16.54
300 1.5	nr	0.42	4.02	13.55	2.64	20.21
450 2.0	nr	0.57	5.45	20.27	3.86	29.58
600 2.0	nr	0.75	7.18	28.73	5.39	41.30

Steel cable tray (cont'd)	Unit	Labour hours	Net labour (£)	Net material (£)	O'heads /profit (£)	Total (£)
Tees						
Width Gauge						
mm mm						
100 1.5	nr	0.30	2.87	4.73	1.14	8.74
150 1.5	nr	0.33	3.16	5.27	1.26	9.69
225 1.5	nr	0.45	4.31	7.55	1.78	13.64
300 1.5	nr	0.62	5.93	11.21	2.57	19.71
450 2.0	nr	0.85	8.13	18.77	4.04	30.94
600 2.0	nr	1.12	10.72	26.51	5.58	42.81
4 way cross piece						
Width Gauge						
mm mm						
100 1.5	nr	0.30	2.87	7.13	1.50	10.00
150 1.5	nr	0.33	3.16	7.55	1.61	10.71
225 1.5	nr	0.45	4.31	10.31	2.19	14.62
300 1.5	nr	0.62	5.93	15.11	3.16	21.04
450 2.0	nr	0.85	8.13	29.27	5.61	43.01
600 2.0	nr	1.12	10.72	39.53	7.54	57.79
Reducers						
Width Gauge						
mm mm						
150 1.5	nr	0.17	1.63	7.91	1.43	10.97
225 1.5	nr	0.23	2.20	8.63	1.62	12.45
300 1.5	nr	0.23	2.20	10.73	1.94	14.87
450 2.0	nr	0.37	3.54	18.41	3.29	25.24
600 2.0	nr	0.37	3.54	23.87	4.11	31.52
Epoxy coated heavy duty cable tray						
Straight tray						
Width Gauge						
mm mm						
100 1.5	m	0.28	2.68	7.30	1.50	11.48
150 1.5	m	0.33	3.16	8.09	1.69	12.94
225 1.5	m	0.40	3.83	9.94	2.07	15.84
300 1.5	m	0.50	4.79	13.06	2.68	20.53
450 2.0	m	0.57	5.45	20.19	3.85	29.49
600 2.0	m	0.80	7.66	25.67	5.00	38.33

M & E MEASUREMENT SERVICES

		Unit	Labour hours	Net labour (£)	Net material (£)	O'heads /profit (£)	Total (£)
Flat bends							
Width	Gauge						
mm	mm						
100	1.5	nr	0.20	1.91	6.53	1.27	9.71
152	1.5	nr	0.25	2.39	7.73	1.52	11.64
225	1.5	nr	0.30	2.87	11.03	2.08	15.98
300	1.5	nr	0.42	4.02	14.09	2.72	20.83
450	2.0	nr	0.57	5.45	25.13	4.59	35.17
600	2.0	nr	0.75	7.18	33.53	6.11	46.82
Outside risers							
Width	Gauge						
mm	mm						
102	1.5	nr	0.20	1.91	14.99	2.54	19.44
150	1.5	nr	0.25	2.39	16.61	2.85	21.85
225	1.5	nr	0.30	2.87	20.39	3.49	26.75
300	1.5	nr	0.42	4.02	25.31	4.40	33.73
450	2.0	nr	0.57	5.45	34.43	5.98	45.86
600	2.0	nr	0.75	7.18	50.57	8.66	66.41
Inside risers							
Width	Gauge						
mm	mm						
100	1.5	nr	0.20	1.91	14.99	2.54	19.44
150	1.5	nr	0.25	2.39	16.61	2.85	21.85
225	1.5	nr	0.30	2.87	20.39	3.49	26.75
300	1.5	nr	0.42	4.02	25.31	4.40	33.73
450	2.0	nr	0.57	5.45	34.43	5.98	45.86
600	2.0	nr	0.75	7.18	50.57	8.66	66.41
Tees							
Width	Gauge						
mm	mm						
100	1.5	nr	0.30	2.87	8.99	1.78	13.64
150	1.5	nr	0.33	3.16	10.01	1.98	15.15
225	1.5	nr	0.45	4.31	14.63	2.84	21.78
300	1.5	nr	0.62	5.93	20.99	4.04	30.96
450	2.0	nr	0.85	8.13	34.07	6.33	48.53
600	2.0	nr	1.12	10.72	47.69	8.76	67.17

CONDUIT AND CABLE TRUNKING

Steel cable tray (cont'd)	Unit	Labour hours	Net labour (£)	Net material (£)	O'heads /profit (£)	Total (£)
4 way cross piece						
Width Gauge						
mm mm						
102 1.5	nr	0.30	2.87	12.47	2.30	17.64
150 1.5	nr	0.33	3.16	13.91	2.56	19.63
225 1.5	nr	0.45	4.31	20.63	3.74	28.68
300 1.5	nr	0.62	5.93	28.79	5.21	39.93
450 2.0	nr	0.85	8.13	48.83	8.54	65.50
600 2.0	nr	1.12	10.72	71.51	12.33	94.56
Reducers						
Width Gauge						
mm mm						
150 1.5	nr	0.17	1.63	16.31	2.69	20.63
225 1.5	nr	0.23	2.20	17.99	3.03	23.22
300 1.5	nr	0.23	2.20	28.79	4.65	35.64
450 2.0	nr	0.37	3.54	36.35	5.98	45.87
600 2.0	nr	0.37	3.54	40.32	6.58	50.44

Hot dip galvanized return flange, medium duty cable tray

Straight tray	Unit	Labour hours	Net labour	Net material	O'heads /profit	Total
Width Gauge						
mm mm						
75 1.5	m	0.25	2.39	3.14	0.83	6.36
102 1.5	m	0.28	2.68	3.47	0.92	7.07
150 1.5	m	0.33	3.16	4.26	1.11	8.53
225 1.5	m	0.40	3.83	4.98	1.32	10.13
300 1.5	m	0.50	4.79	6.65	1.72	13.16
450 2.0	m	0.57	5.45	10.59	2.41	18.45
600 2.0	m	0.80	7.66	14.14	3.27	25.07
Flat bends						
Width Gauge						
mm mm						
75 1.5	nr	0.18	1.72	9.59	1.70	13.01
100 1.5	nr	0.20	1.91	10.97	1.93	14.81
150 1.5	nr	0.25	2.39	11.81	2.13	16.33
225 1.5	nr	0.30	2.87	14.09	2.54	19.50

		Unit	Labour hours	Net labour (£)	Net material (£)	O'heads /profit (£)	Total (£)
300	1.5	nr	0.42	4.02	18.05	3.31	25.38
450	2.0	nr	0.57	5.45	28.55	5.10	39.10
600	2.0	nr	0.75	7.18	35.09	6.34	48.61

Outside risers
Width Gauge
mm   mm

75	1.5	nr	0.18	1.72	6.11	1.17	7.83
100	1.5	nr	0.20	1.91	6.23	1.22	9.36
150	1.5	nr	0.25	2.39	7.67	1.51	11.57
225	1.5	nr	0.30	2.87	8.81	1.75	13.43
300	1.5	nr	0.42	4.02	9.83	2.08	15.93
450	2.0	nr	0.57	5.45	15.41	3.13	23.99
600	2.0	nr	0.75	7.18	20.81	4.20	32.19

Inside risers
Width Gauge
mm   mm

75	1.5	nr	0.18	1.72	6.11	1.17	9.00
100	1.5	nr	0.20	1.91	6.23	1.22	9.36
150	1.5	nr	0.25	2.39	7.67	1.51	11.57
225	1.5	nr	0.30	2.87	8.81	1.75	13.43
300	1.5	nr	0.42	4.02	9.83	2.08	15.93
450	2.0	nr	0.57	5.45	15.41	3.13	23.99
600	2.0	nr	0.75	7.18	20.81	4.20	32.19

Tees
Width Gauge
mm   mm

75	1.5	nr	0.18	1.72	13.25	2.25	17.22
100	1.5	nr	0.30	2.87	13.37	2.44	18.68
150	1.5	nr	0.33	3.16	15.89	2.86	21.91
225	1.5	nr	0.45	4.31	17.57	3.28	25.16
300	1.5	nr	0.62	5.93	22.31	4.24	32.48
450	2.0	nr	0.85	8.13	30.83	5.84	44.80
600	2.0	nr	1.12	10.72	46.25	8.55	65.52

CONDUIT AND CABLE TRUNKING

Steel cable tray (cont'd)	Unit	Labour hours	Net labour (£)	Net material (£)	O'heads /profit (£)	Total (£)
4 way cross piece						
Width Gauge						
mm mm						
75 1.5	nr	0.28	2.68	19.01	3.25	24.94
100 1.5	nr	0.30	2.87	20.99	3.58	27.44
150 1.5	nr	0.33	3.16	22.49	3.85	29.50
225 1.5	nr	0.45	4.31	27.89	4.83	37.03
300 1.5	nr	0.62	5.93	33.47	5.91	45.31
450 2.0	nr	0.85	8.13	44.27	7.86	60.26
600 2.0	nr	1.12	10.72	67.25	11.70	89.67
Reducers						
Width Gauge						
mm mm						
100 1.5	nr	0.14	1.34	5.93	1.09	7.27
150 1.5	nr	0.17	1.63	6.77	1.26	9.66
225 1.5	nr	0.23	2.20	8.15	1.55	11.90
300 1.5	nr	0.23	2.20	9.83	1.80	13.83
450 2.0	nr	0.37	3.54	13.49	2.55	19.58
600 2.0	nr	0.37	3.54	17.39	3.14	24.07

Epoxy coated, return flange
medium duty cable tray

Straight tray						
Width Gauge						
mm mm						
75 1.5	m	0.20	1.91	5.13	1.06	8.10
100 1.5	m	0.26	2.49	5.60	1.21	9.30
150 1.5	m	0.33	3.16	7.67	1.62	12.45
225 1.5	m	0.40	3.83	9.35	1.98	15.16
300 1.5	m	0.50	4.79	12.30	2.56	19.65
450 2.0	m	0.57	5.45	17.35	3.42	26.22
600 2.0	m	0.80	7.66	23.62	4.69	35.97

M & E MEASUREMENT SERVICES

		Unit	Labour hours	Net labour (£)	Net material (£)	O'heads /profit (£)	Total (£)
Flat bends							
Width	Gauge						
mm	mm						
75	1.5	nr	0.18	1.72	14.03	2.36	18.11
100	1.5	nr	0.20	1.91	14.39	2.44	18.74
150	1.5	nr	0.25	2.39	17.75	3.02	23.16
225	1.5	nr	0.30	2.87	21.53	3.66	28.06
300	1.5	nr	0.42	4.02	26.39	4.56	34.97
450	2.0	nr	0.57	5.45	33.47	5.84	44.76
600	2.0	nr	0.75	7.18	50.81	8.70	66.69
Outside risers							
Width	Gauge						
mm	mm						
75	1.5	nr	0.18	1.72	8.63	1.55	11.90
100	1.5	nr	0.20	1.91	8.93	1.63	12.47
150	1.5	nr	0.25	2.39	14.27	2.50	19.16
225	1.5	nr	0.30	2.87	17.63	3.07	23.57
300	1.5	nr	0.42	4.02	23.63	4.15	31.80
450	2.0	nr	0.57	5.45	30.29	5.36	41.10
600	2.0	nr	0.75	7.18	39.23	6.96	53.37
Inside risers							
Width	Gauge						
mm	mm						
75	1.5	nr	0.18	1.72	8.63	1.55	11.90
100	1.5	nr	0.20	1.91	8.93	1.63	12.47
150	1.5	nr	0.25	2.39	14.27	2.50	19.16
225	1.5	nr	0.30	2.87	17.63	3.07	23.57
300	1.5	nr	0.42	4.02	23.63	4.15	31.80
450	2.0	nr	0.57	5.45	30.29	5.36	41.10
600	2.0	nr	0.75	7.18	39.23	6.96	53.37
Tees							
Width	Gauge						
mm	mm						
75	1.5	nr	0.28	2.68	18.83	3.23	24.74
100	1.5	nr	0.30	2.87	19.49	3.35	25.71
150	1.5	nr	0.33	3.16	23.81	4.05	31.02
225	1.5	nr	0.45	4.31	28.19	4.88	37.38

CONDUIT AND CABLE TRUNKING

Steel cable tray (cont'd)		Unit	Labour hours	Net labour (£)	Net material (£)	O'heads /profit (£)	Total (£)
300	1.5	nr	0.62	5.93	32.82	5.81	44.56
450	2.0	nr	0.85	8.13	46.91	8.26	63.30
600	2.0	nr	1.12	10.72	73.43	12.62	96.77
4 way cross piece							
Width	Gauge						
mm	mm						
75	1.5	nr	0.28	2.68	30.95	5.04	38.67
100	1.5	nr	0.30	2.87	31.43	5.14	34.30
150	1.5	nr	0.33	3.16	38.69	6.28	48.13
225	1.5	nr	0.45	4.31	44.33	7.30	55.94
300	1.5	nr	0.62	5.93	51.65	8.64	66.22
450	2.0	nr	0.85	8.13	70.61	11.81	90.55
600	2.0	nr	1.12	10.72	109.97	18.10	138.79
Reducers							
Width	Gauge						
mm	mm						
100	1.5	nr	0.17	1.63	11.21	1.93	14.77
150	1.5	nr	0.17	1.63	11.57	1.98	15.18
225	1.5	nr	0.23	2.20	15.11	2.60	19.91
300	1.5	nr	0.23	2.20	19.25	3.22	24.67
450	2.0	nr	0.37	3.54	27.47	4.65	35.66
600	2.0	nr	0.37	3.54	37.73	6.19	47.46
Pre-galvanized return flange, medium duty cable tray							
Straight tray							
Width	Gauge						
mm	mm						
75	1.5	m	0.20	1.91	2.34	0.64	4.89
100	1.5	m	0.25	2.39	2.59	0.75	5.73
150	1.5	m	0.28	2.68	3.14	0.87	6.69
225	1.5	m	0.30	2.87	3.93	1.02	7.82
300	1.5	m	0.42	4.02	6.29	1.55	11.86
450	2.0	m	0.57	5.45	8.38	2.07	15.90
600	2.0	m	0.75	7.18	10.75	2.69	20.62

M & E MEASUREMENT SERVICES

		Unit	Labour hours	Net labour (£)	Net material (£)	O'heads /profit (£)	Total (£)
Flat bends							
Width	Gauge						
mm	mm						
450	2.0	nr	0.57	5.45	22.19	4.15	31.79
600	2.0	nr	0.75	7.18	29.33	5.48	41.99
Tees							
Width	Gauge						
mm	mm						
75	1.5	nr	0.28	2.68	12.89	2.34	17.91
100	1.5	nr	0.30	2.87	12.95	2.37	18.19
150	1.5	nr	0.33	3.16	14.51	2.65	20.32
225	1.5	nr	0.45	4.31	16.07	3.06	23.44
300	1.5	nr	0.62	5.93	20.69	3.99	30.61
450	2.0	nr	0.85	8.13	28.07	5.43	41.63
600	2.0	nr	1.12	10.72	41.87	7.89	60.48
4 way cross piece							
Width	Gauge						
mm	mm						
75	1.5	nr	0.28	2.68	17.51	3.03	23.22
100	1.5	nr	0.30	2.87	19.67	3.38	25.92
150	1.5	nr	0.33	3.16	20.69	3.58	27.43
225	1.5	nr	0.45	4.31	25.91	4.53	34.75
300	1.5	nr	0.62	5.93	30.17	5.42	41.52
450	2.0	nr	0.85	8.13	41.21	7.40	56.74
600	2.0	nr	1.12	10.72	63.23	11.09	85.04
Reducers							
Width	Gauge						
mm	mm						
100	1.5	nr	0.14	1.34	5.51	1.03	7.88
150	1.5	nr	0.17	1.63	5.81	1.12	8.56
225	1.5	nr	0.23	2.20	7.31	1.43	10.94
300	1.5	nr	0.23	2.20	9.59	1.77	13.56
450	2.0	nr	0.37	3.54	13.13	2.50	19.17
600	2.0	nr	0.37	3.54	16.13	2.95	22.62

Steel cable tray (cont'd)	Unit	Labour hours	Net labour (£)	Net material (£)	O'heads /profit (£)	Total (£)
Hot dip galvanized return flange, heavy duty cable tray						
Straight tray						
Width Gauge						
mm mm						
75 1.5	m	0.25	2.39	5.92	1.25	9.56
100 1.5	m	0.28	2.68	6.35	1.35	10.38
150 1.5	m	0.33	3.16	7.28	1.57	12.01
225 1.5	m	0.40	3.83	8.16	1.80	13.79
300 1.5	m	0.50	4.79	9.90	2.20	16.89
450 2.0	m	0.67	6.41	14.94	3.20	24.55
600 2.0	m	0.80	7.66	18.10	3.86	29.62
750 2.0	m	0.89	8.52	22.79	4.70	36.01
900 2.0	m	0.98	9.38	25.23	5.19	39.80
Flat bends						
Width Gauge						
mm mm						
75 1.5	nr	0.18	1.72	13.07	2.22	17.01
100 1.5	nr	0.20	1.91	15.41	2.60	19.92
150 1.5	nr	0.25	2.39	16.73	2.87	21.99
225 1.5	nr	0.30	2.87	19.37	3.34	25.58
300 1.5	nr	0.42	4.02	22.67	4.00	30.69
450 2.0	nr	0.57	5.45	35.09	6.08	46.62
600 2.0	nr	0.75	7.18	48.23	8.31	63.72
750 2.0	nr	0.87	8.33	69.17	11.62	89.12
900 2.0	nr	0.96	9.19	79.01	13.23	101.43
Outside risers						
Width Gauge						
mm mm						
75 1.5	nr	0.18	1.72	9.59	1.70	13.01
100 1.5	nr	0.20	1.91	10.25	1.82	13.98
150 1.5	nr	0.25	2.39	12.17	2.18	16.74
225 1.5	nr	0.30	2.87	13.79	2.50	19.16
300 1.5	nr	0.42	4.02	15.41	2.91	22.34
450 2.0	nr	0.57	5.45	20.69	3.92	30.06
600 2.0	nr	0.75	7.18	26.27	5.02	38.47
750 2.0	nr	0.87	8.33	36.41	6.71	51.45
900 2.0	nr	0.96	9.19	43.61	7.92	60.72

M & E MEASUREMENT SERVICES

		Unit	Labour hours	Net labour (£)	Net material (£)	O'heads /profit (£)	Total (£)
Inside risers							
Width	Gauge						
mm	mm						
75	1.5	nr	0.18	1.72	9.59	1.70	13.01
100	1.5	nr	0.20	1.91	10.25	1.82	13.98
150	1.5	nr	0.25	2.39	12.17	2.18	16.74
225	1.5	nr	0.30	2.87	13.79	2.50	19.16
300	1.5	nr	0.42	4.02	15.41	2.91	22.34
450	2.0	nr	0.57	5.45	20.69	3.92	30.06
600	2.0	nr	0.75	7.18	26.27	5.02	38.47
750	2.0	nr	0.87	8.33	36.41	6.71	51.45
600	2.0	nr	0.96	9.19	43.61	7.92	60.72
Tees							
Width	Gauge						
mm	mm						
75	1.5	nr	0.23	2.20	16.61	2.82	21.63
100	1.5	nr	0.30	2.87	19.25	3.32	25.44
150	1.5	nr	0.33	3.16	21.65	3.72	28.53
225	1.5	nr	0.45	4.31	25.91	4.53	34.75
300	1.5	nr	0.62	5.93	28.55	7.53	34.48
450	2.0	nr	0.85	8.13	44.27	10.71	52.40
600	2.0	nr	1.12	10.72	63.29	14.45	74.01
750	2.0	nr	1.32	12.63	85.61	14.74	112.98
900	2.0	nr	1.49	14.26	98.69	16.94	129.89
4 way cross piece							
Width	Gauge						
mm	mm						
75	1.5	nr	0.28	2.68	23.63	3.95	30.26
100	1.5	nr	0.30	2.87	24.77	4.15	31.79
150	1.5	nr	0.33	3.16	32.15	5.30	40.61
225	1.5	nr	0.45	4.31	37.73	6.31	48.35
300	1.5	nr	0.62	5.93	42.95	7.33	56.21
450	2.0	nr	0.85	8.13	67.91	11.41	87.45
600	2.0	nr	1.12	10.72	95.45	15.93	122.10
750	2.0	nr	1.32	12.63	118.37	19.65	150.65
900	2.0	nr	1.49	14.26	144.59	23.83	182.68

CONDUIT AND CABLE TRUNKING

Steel cable tray (cont'd)	Unit	Labour hours	Net labour (£)	Net material (£)	O'heads /profit (£)	Total (£)
Reducers						
Width Gauge						
mm mm						
100 1.5	nr	0.14	1.34	10.19	1.73	13.26
150 1.5	nr	0.17	1.63	10.85	1.87	14.35
225 1.5	nr	0.23	2.20	12.83	2.25	17.28
300 1.5	nr	0.23	2.20	14.75	2.54	19.49
450 2.0	nr	0.37	3.54	22.67	3.93	30.14
600 2.0	nr	0.37	3.54	27.89	4.71	36.14
750 2.0	nr	0.48	4.59	36.41	6.15	47.15
900 2.0	nr	0.54	5.17	40.37	6.83	52.37

Epoxy coated return flange, heavy duty cable tray

Straight tray						
Width Gauge						
mm mm						
75 1.5	nr	0.25	2.39	8.72	1.67	12.78
100 1.5	nr	0.28	2.68	9.50	1.83	14.01
150 1.5	nr	0.33	3.16	11.32	2.17	16.65
225 1.5	nr	0.40	3.83	13.66	2.62	20.11
300 1.5	nr	0.50	4.79	16.26	3.16	24.21
450 2.0	nr	0.57	5.45	24.70	4.52	34.67
600 2.0	nr	0.80	7.66	30.22	5.68	43.56
750 2.0	nr	0.89	8.52	37.57	6.91	53.00
900 2.0	nr	0.98	9.38	44.96	8.15	62.49

Flat bends						
Width Gauge						
mm mm						
75 1.5	nr	0.18	1.72	19.61	3.20	24.53
100 1.5	nr	0.20	1.91	20.09	3.30	25.30
150 1.5	nr	0.25	2.39	22.37	3.71	28.47
225 1.5	nr	0.30	2.87	28.49	4.70	36.06
300 1.5	nr	0.42	4.02	33.65	5.65	43.32
450 2.0	nr	0.57	5.45	55.19	9.10	69.74
600 2.0	nr	0.75	7.18	73.19	12.06	92.43
750 2.0	nr	0.87	8.33	91.43	14.96	114.72
900 2.0	nr	0.96	9.19	123.89	19.96	153.04

M & E MEASUREMENT SERVICES

	Unit	Labour hours	Net labour (£)	Net material (£)	O'heads /profit (£)	Total (£)
Outside risers						
Width Gauge						
mm mm						
75 1.5	nr	0.18	1.72	17.93	2.95	22.60
100 1.5	nr	0.20	1.91	18.71	3.09	23.71
150 1.5	nr	0.25	2.39	20.75	3.47	26.61
225 1.5	nr	0.30	2.87	25.37	4.24	32.48
300 1.5	nr	0.42	4.02	28.43	4.87	37.32
450 2.0	nr	0.57	5.45	40.67	6.92	53.04
600 2.0	nr	0.75	7.18	50.09	8.59	65.86
750 2.0	nr	0.87	8.33	62.57	10.63	81.53
900 2.0	nr	0.96	9.19	74.51	12.55	96.25
Inside risers						
Width Gauge						
mm mm						
75 1.5	nr	0.18	1.72	17.93	2.95	22.60
100 1.5	nr	0.20	1.91	18.71	3.09	23.71
150 1.5	nr	0.25	2.39	20.75	3.47	26.61
225 1.5	nr	0.30	2.87	25.37	4.24	32.48
300 1.5	nr	0.42	4.02	28.43	4.87	37.32
450 2.0	nr	0.57	5.45	40.67	6.92	53.04
600 2.0	nr	0.75	7.18	50.09	8.59	65.86
750 2.0	nr	0.87	8.33	62.57	10.63	81.53
900 2.0	nr	0.96	9.19	74.51	12.55	96.25
Tees						
Width Gauge						
mm mm						
75 1.5	nr	0.28	2.68	23.45	3.92	30.05
100 1.5	nr	0.30	2.87	26.27	4.37	33.51
150 1.5	nr	0.33	3.16	29.15	4.85	37.16
225 1.5	nr	0.45	4.31	37.67	6.30	48.28
300 1.5	nr	0.62	5.93	43.37	7.39	56.69
450 2.0	nr	0.85	8.13	70.07	11.73	89.93
600 2.0	nr	1.12	10.72	98.63	16.40	125.75
750 2.0	nr	1.32	12.63	123.22	20.38	156.23
900 2.0	nr	1.49	14.26	148.55	24.42	187.23

CONDUIT AND CABLE TRUNKING

Steel cable tray (cont'd)	Unit	Labour hours	Net labour (£)	Net material (£)	O'heads /profit (£)	Total (£)
4 way cross piece						
Width Gauge						
mm mm						
75 1.5	nr	0.28	2.68	35.09	5.67	43.44
100 1.5	nr	0.30	2.87	35.57	5.77	44.21
150 1.5	nr	0.33	3.16	43.37	6.98	53.51
225 1.5	nr	0.45	4.31	56.39	9.11	69.81
300 1.5	nr	0.62	5.93	64.85	10.62	81.40
450 2.0	nr	0.85	8.13	104.87	16.95	129.95
600 2.0	nr	1.12	10.72	140.27	22.65	173.64
250 2.0	nr	1.32	12.63	175.31	28.19	216.13
900 2.0	nr	1.49	14.26	210.29	33.68	258.23
Reducers						
Width Gauge						
mm mm						
100 1.5	nr	0.14	1.34	16.85	2.73	20.92
150 1.5	nr	0.17	1.63	17.99	2.94	22.56
225 1.5	nr	0.23	2.20	21.83	3.60	27.63
300 1.5	nr	0.23	2.20	26.39	4.29	32.88
450 2.0	nr	0.37	3.54	39.35	6.43	49.32
600 2.0	nr	0.37	3.54	41.09	6.69	51.32
750 2.0	nr	0.48	4.59	43.13	7.16	54.88
900 2.0	nr	0.54	5.17	45.17	7.55	57.89
Pre-galvanized return flange, heavy duty cable tray						
Straight tray						
Width Gauge						
mm mm						
75 1.5	nr	0.25	2.39	4.70	1.06	7.09
100 1.5	nr	0.28	2.68	5.25	1.19	9.12
150 1.5	nr	0.33	3.16	5.76	1.34	10.26
225 1.5	nr	0.40	3.83	6.33	1.52	11.68
300 1.5	nr	0.50	4.79	6.73	1.73	13.25
450 2.0	nr	0.57	5.45	11.18	2.49	19.12
600 2.0	nr	0.80	7.66	15.44	3.46	26.56
750 2.0	nr	0.89	8.52	19.71	4.23	28.23
900 2.0	nr	0.98	9.38	20.62	4.50	30.00

M & E MEASUREMENT SERVICES

		Unit	Labour hours	Net labour (£)	Net material (£)	O'heads /profit (£)	Total (£)
Flat bends							
Width	Gauge						
mm	mm						
450	2.0	nr	0.57	5.45	31.25	5.50	42.20
600	2.0	nr	0.75	7.18	43.85	7.65	58.68
750	2.0	nr	0.87	8.33	56.45	9.72	74.50
900	2.0	nr	0.96	9.19	59.15	10.25	78.59
Tees							
Width	Gauge						
mm	mm						
75	1.5	nr	0.28	2.68	15.41	2.71	20.80
100	1.5	nr	0.30	2.87	16.91	2.97	22.75
150	1.5	nr	0.33	3.16	18.89	3.31	25.36
225	1.5	nr	0.45	4.31	21.59	3.88	29.78
300	1.5	nr	0.62	5.93	23.57	4.42	33.92
450	2.0	nr	0.85	8.13	36.95	6.76	51.84
600	2.0	nr	1.12	10.72	52.67	9.51	72.90
750	2.0	nr	1.32	12.63	65.57	11.73	89.93
900	2.0	nr	1.49	14.26	81.29	14.33	109.88
4 way cross piece							
Width	Gauge						
mm	mm						
75	1.5	nr	0.28	2.68	22.43	3.77	28.88
100	1.5	nr	0.30	2.87	24.23	4.06	31.16
150	1.5	nr	0.33	3.16	27.65	4.62	35.43
225	1.5	nr	0.45	4.31	32.09	5.46	41.86
300	1.5	nr	0.62	5.93	35.57	6.22	47.72
450	2.0	nr	0.85	8.13	54.89	9.45	72.47
600	2.0	nr	1.12	10.72	72.95	12.55	96.22
750	2.0	nr	1.32	12.63	87.35	15.00	114.98
900	2.0	nr	1.49	14.26	92.81	16.06	123.13
Reducers							
Width	Gauge						
mm	mm						
100	1.5	nr	0.14	1.34	7.85	1.38	10.57
150	1.5	nr	0.17	1.63	7.91	1.43	10.97
225	1.5	nr	0.23	2.20	10.07	1.84	14.11

Steel cable tray (cont'd)	Unit	Labour hours	Net labour (£)	Net material (£)	O'heads /profit (£)	Total (£)
300 1.5	nr	0.23	2.20	11.21	2.01	15.42
450 2.0	nr	0.37	3.54	16.55	3.01	23.10
600 2.0	nr	0.37	3.54	17.93	3.22	24.69
750 2.0	nr	0.48	4.59	21.95	3.98	30.52
900 2.0	nr	0.54	5.17	23.57	4.31	33.05

Hot dip galvanized, return flange,
extra heavy duty cable tray

Straight tray
Width Gauge
mm   mm

150 1.5	nr	0.33	3.16	9.97	1.97	15.10
225 1.5	nr	0.40	3.83	11.29	2.27	17.39
300 1.5	nr	0.50	4.79	12.86	2.65	20.30
450 2.0	nr	0.57	5.45	18.90	3.65	28.00
600 2.0	nr	0.80	7.66	22.40	4.51	34.57

Flat bends
Width Gauge
mm   mm

150 1.5	nr	0.25	2.39	22.85	3.79	29.03
225 1.5	nr	0.30	2.87	26.81	4.45	34.13
300 1.5	nr	0.42	4.02	32.57	5.49	42.08
450 2.0	nr	0.57	5.45	41.39	7.03	53.87
600 2.0	nr	0.75	7.18	68.87	11.41	87.46

Outside risers
Width Gauge
mm   mm

150 1.5	nr	0.25	2.39	17.03	2.91	22.33
225 1.5	nr	0.30	2.87	19.91	3.42	26.20
300 1.5	nr	0.42	4.02	23.51	4.13	31.66
450 2.0	nr	0.57	5.45	32.75	5.73	43.93
600 2.0	nr	0.75	7.18	39.29	6.97	53.44

M & E MEASUREMENT SERVICES

		Unit	Labour hours	Net labour (£)	Net material (£)	O'heads /profit (£)	Total (£)
Inside risers							
Width	Gauge						
mm	mm						
150	1.5	nr	0.25	2.39	17.03	2.91	22.33
225	1.5	nr	0.30	2.87	19.91	3.42	22.78
300	1.5	nr	0.42	4.02	23.51	4.13	31.66
450	2.0	nr	0.57	5.45	32.75	5.73	43.93
600	2.0	nr	0.75	7.18	39.29	6.97	53.44
Tees							
Width	Gauge						
mm	mm						
150	1.5	nr	0.33	3.16	29.27	4.86	37.29
225	1.5	nr	0.45	4.31	35.27	5.94	45.52
300	1.5	nr	0.62	5.93	41.09	7.05	54.07
450	2.0	nr	0.85	8.13	62.69	10.62	81.44
600	2.0	nr	1.12	10.72	88.43	14.87	114.02
4 way cross piece							
Width	Gauge						
mm	mm						
150	1.5	nr	0.33	3.16	43.55	7.01	53.72
225	1.5	nr	0.45	4.31	52.91	8.58	65.80
300	1.5	nr	0.62	5.93	61.31	10.09	77.33
450	2.0	nr	0.85	8.13	94.01	15.32	117.46
600	2.0	nr	1.12	10.72	132.59	21.50	164.81
Reducers							
Width	Gauge						
mm	mm						
225	1.5	nr	0.23	2.20	17.15	2.90	19.35
305	1.5	nr	0.23	2.20	20.21	3.36	25.77
450	2.0	nr	0.37	3.54	30.95	5.17	39.66
600	2.0	nr	0.37	3.54	37.31	6.13	46.98

Steel cable tray (cont'd)	Unit	Labour hours	Net labour (£)	Net material (£)	O'heads /profit (£)	Total (£)
Fish plates, hot dip galvanized, tray width						
75mm	nr	0.12	1.15	0.58	0.26	1.73
102mm	nr	0.15	1.44	0.58	0.30	2.02
150mm	nr	0.27	2.58	0.81	0.51	3.39
Fish plates, epoxy coated, tray width						
75mm	nr	0.13	1.24	1.38	0.39	3.01
102mm	nr	0.13	1.24	1.61	0.43	3.28
150mm	nr	0.13	1.24	2.19	0.51	3.94

Channel support system for cable tray and trunking in 3m lengths. A 10% waste factor has been added to the material costs of the channels

Galvanized finish to BS2989

	Unit	Labour hours	Net labour (£)	Net material (£)	O'heads /profit (£)	Total (£)
shallow channel section	nr	0.40	3.83	9.16	1.95	14.94
deep channel section	nr	4.00	38.28	11.14	7.41	56.83
slotted shallow channel section	nr	0.40	3.83	9.65	2.02	13.48
slotted deep channel section	nr	0.40	3.83	11.88	2.36	15.71
back to back shallow channel section	nr	0.40	3.83	21.14	3.75	28.72
back to back deep channel section	nr	0.40	3.83	24.75	4.29	32.87

Hot dip galvanized finish to BS72

	Unit	Labour hours	Net labour (£)	Net material (£)	O'heads /profit (£)	Total (£)
shallow channel section	nr	0.40	3.83	14.65	2.77	21.25
deep channel section	nr	0.40	3.83	13.86	2.65	20.34
back to back shallow channel section	nr	0.40	3.83	24.75	4.29	32.87
back to back deep channel section	nr	0.40	3.83	31.68	5.33	40.84

	Unit	Labour hours	Net labour (£)	Net material (£)	O'heads /profit (£)	Total (£)
Concrete inserts						
Galvanized to BS2989						
3m shallow insert	nr	0.20	1.91	12.60	2.18	16.69
3m deep insert	nr	0.20	1.91	16.20	2.72	20.83
Hot dip galvanized to BS2729						
3m shallow insert	nr	0.20	1.91	17.10	2.85	21.86
3m deep insert	nr	0.20	1.91	20.70	3.39	26.00
Channel nuts (in pack of 100 only)						
Long spring						
M6	100	1.20	11.48	25.65	5.57	42.70
M8	100	1.20	11.48	33.75	6.78	52.01
M10	100	1.20	11.48	34.20	6.85	52.53
M12	100	1.20	11.48	37.35	7.32	56.15

A discount of 10% has been
incorporated within the net
material costs of these items

	Unit	Labour hours	Net labour (£)	Net material (£)	O'heads /profit (£)	Total (£)
Short spring						
M6	100	1.20	11.48	26.55	5.70	43.73
M8	100	1.20	11.48	32.85	6.65	50.98
M10	100	1.20	11.48	35.55	7.05	54.08
No spring						
M6	100	1.20	11.48	18.00	4.42	33.90
M8	100	1.20	11.48	22.50	5.10	39.08
M10	100	1.20	11.48	24.30	5.37	41.15
M12	100	1.20	11.48	32.40	6.58	50.46
Hexagon head set screws (in packs of 200 only)						
M6 x 20mm	100	0.80	7.66	6.40	2.11	16.17
M6 x 25mm	100	0.80	7.66	6.40	2.11	16.17
M8 x 25mm	100	0.80	7.66	9.23	2.53	19.42

Steel cable tray (cont'd)	Unit	Labour hours	Net labour (£)	Net material (£)	O'heads /profit (£)	Total (£)
M10 x 16mm	100	0.80	7.66	11.33	2.85	21.84
M10 x 20mm	100	0.80	7.66	10.46	2.72	20.84
M10 x 25mm	100	0.80	7.66	10.46	2.72	20.84
(in packs of 100 only)						
M10 x 35mm	100	0.80	7.66	13.15	3.12	23.93
M10 x 50mm	100	0.80	7.66	17.90	3.83	29.39
M12 x 20mm	100	0.80	7.66	20.19	4.18	32.03
M12 x 25mm	100	0.80	7.66	20.19	4.18	32.03
M12 x 35mm	100	0.80	7.66	23.57	4.68	35.91
Other fasteners						
M6 HT hexagon nut (500 in pack)	100	0.80	7.66	2.61	1.54	11.81
M8 HT hexagon nut (500 in pack)	100	0.80	7.66	4.49	1.82	13.97
M10 HT hexagon nut (200 in pack)	100	0.80	7.66	5.53	1.98	15.17
M12 HT hexagon nut (200 in pack)	100	0.80	7.66	10.69	2.75	21.10
M6 flat washer (500 in pack)	100	0.40	3.83	1.77	0.84	6.44
M8 flat washer (500 in pack)	100	0.40	3.83	2.08	0.89	6.80
M10 flat washer (500 in pack)	100	0.40	3.83	2.45	0.94	7.22
M12 flat washer (200 in pack)	100	0.40	3.83	5.76	1.44	11.03
M6 x 16mm galvanized roofing bolts and nuts (200 in pack)	100	1.00	9.57	6.48	2.41	18.46
M6 x 3m threaded rod	m	0.17	1.63	2.74	0.66	5.03
M10 x 3m threaded rod	m	0.17	1.63	3.65	0.79	6.07
M12 x 3m threaded rod	m	0.17	1.63	5.48	1.07	8.18
M6 x 18mm threaded rod connector	nr	0.03	0.29	0.90	0.18	1.37
M10 x 30mm threaded rod connector	nr	0.03	0.29	0.99	0.19	1.47
M12 x 36mm threaded rod connector	nr	0.03	0.29	1.26	0.23	1.78
M6 x 80mm eye bolt	nr	0.12	1.15	0.27	0.21	1.63
M10 x 80mm eye bolt	nr	0.12	1.15	0.36	0.23	1.74
M12 x 80mm eye bolt	nr	0.12	1.15	0.81	0.29	2.25

	Unit	Labour hours	Net labour (£)	Net material (£)	O'heads /profit (£)	Total (£)
Cable ladder systems galvanized to BS729						
A discount of 40% has been incorporated within the net material costs in this section						
Light duty, 60mm deep, gauge 1.5mm						
ladder 3m length, width						
150mm	m	0.33	3.16	5.67	1.32	10.15
300mm	m	0.57	5.45	6.39	1.78	13.62
450mm	m	0.57	5.45	7.36	1.92	14.73
600mm	m	0.80	7.66	8.05	2.36	18.07
750mm	m	0.87	8.33	8.90	2.58	19.81
900mm	m	0.98	9.38	9.92	2.90	22.20
internal riser, width						
150mm	nr	0.33	3.16	18.06	3.18	24.40
300mm	nr	0.67	6.41	18.99	3.81	29.21
450mm	nr	0.83	7.94	20.39	4.25	28.33
600mm	nr	1.03	9.86	21.15	4.65	35.66
750mm	nr	1.10	10.53	22.39	4.94	37.86
900mm	nr	1.17	11.20	23.34	5.18	39.72
external riser, width						
150mm	nr	0.33	3.16	18.97	3.32	25.45
300mm	nr	0.67	6.41	19.92	3.95	30.28
450mm	nr	0.83	7.94	20.77	4.31	33.02
600mm	nr	1.03	9.86	21.52	4.71	36.09
750mm	nr	1.10	10.53	22.76	4.99	38.28
900mm	nr	1.17	11.20	23.70	5.23	40.13
90 degrees flat bend, width						
150mm	nr	11.35	108.62	11.35	18.00	137.97
300mm	nr	0.33	3.16	12.77	2.39	18.32
450mm	nr	0.83	7.94	14.05	3.30	25.29
600mm	nr	1.00	9.57	16.74	3.95	30.26
750mm	nr	1.10	10.53	16.74	4.09	31.36
900mm	nr	1.20	11.48	18.29	4.47	34.24

CONDUIT AND CABLE TRUNKING

Steel cable tray (cont'd)	Unit	Labour hours	Net labour (£)	Net material (£)	O'heads /profit (£)	Total (£)
45 degrees flat bend, width						
150mm	nr	0.27	2.58	9.00	1.74	13.32
300mm	nr	0.33	3.16	10.16	2.00	15.32
450mm	nr	0.83	7.94	10.97	2.84	21.75
600mm	nr	1.00	9.57	11.76	3.20	24.53
750mm	nr	1.10	10.53	13.59	3.62	27.74
900mm	nr	1.20	11.48	14.75	3.93	30.16
equal tee, width	nr	0.37	3.54	19.54	3.46	23.08
300mm	nr	0.57	5.45	23.05	4.28	32.78
450mm	nr	0.97	9.28	27.13	5.46	41.87
600mm	nr	1.17	11.20	28.82	6.00	46.02
750mm	nr	1.25	11.96	33.78	6.86	52.60
900mm	nr	1.33	12.73	37.33	7.51	57.57
four way, width						
150mm	nr	0.50	4.79	29.26	5.11	39.16
300mm	nr	0.67	6.41	31.14	5.63	43.18
450mm	nr	1.17	11.20	33.97	6.78	51.95
600mm	nr	1.37	13.11	35.81	7.34	56.26
750mm	nr	1.43	13.69	39.62	8.00	61.31
900mm	nr	1.50	14.36	42.46	8.52	65.34
straight reducer, width						
300mm	nr	0.27	2.58	12.09	2.20	16.87
450mm	nr	0.27	2.58	12.65	2.28	17.51
600mm	nr	0.27	2.58	13.15	2.36	18.09
750mm	nr	0.27	2.58	15.33	2.69	20.60
900mm	nr	0.27	2.58	17.53	3.02	20.11
offset reducer, width						
300mm	nr	0.27	2.58	9.07	1.75	13.40
450mm	nr	0.27	2.58	9.69	1.84	14.11
600mm	nr	0.27	2.58	10.18	1.91	14.67
750mm	nr	0.27	2.58	12.16	2.21	16.95
900mm	nr	0.27	2.58	14.31	2.53	19.42
couplers (pairs) all sizes	nr	0.54	5.17	1.71	1.03	7.91

M & E MEASUREMENT SERVICES

	Unit	Labour hours	Net labour (£)	Net material (£)	O'heads /profit (£)	Total (£)
Medium duty, 90mm deep, gauge 1.5mm						
ladder 3m length, width						
150mm	m	0.33	3.16	9.39	1.88	14.43
300mm	m	0.57	5.45	10.21	2.35	18.01
450mm	m	0.57	5.45	11.29	2.51	19.25
600mm	m	0.80	7.66	12.07	2.96	22.69
750mm	m	0.87	8.33	13.54	3.28	25.15
900mm	m	0.98	9.38	14.53	3.59	27.50
internal riser, width						
150mm	nr	0.33	3.16	24.20	4.10	31.46
300mm	nr	0.67	6.41	24.46	4.63	35.50
450mm	nr	0.83	7.94	26.61	5.18	39.73
600mm	nr	1.03	9.86	29.00	5.83	44.69
750mm	nr	1.10	10.53	32.17	6.41	49.11
900mm	nr	1.17	11.20	34.20	6.81	52.21
external riser, width						
150mm	nr	0.33	3.16	24.20	4.10	31.46
300mm	nr	0.67	6.41	24.46	4.63	35.50
450mm	nr	0.83	7.94	26.61	5.18	39.73
600mm	nr	1.03	9.86	29.00	5.83	44.69
750mm	nr	1.10	10.53	32.17	6.41	49.11
900mm	nr	1.17	11.20	34.93	6.92	53.05
90 degrees flat bend, width						
150mm	nr	0.27	2.58	15.59	2.73	20.90
300mm	nr	0.33	3.16	16.44	2.94	22.54
450mm	nr	0.83	7.94	19.16	4.06	31.16
600mm	nr	1.00	9.57	22.57	4.82	36.96
750mm	nr	1.10	10.53	25.59	5.42	41.54
900mm	nr	1.20	11.48	28.46	5.99	45.93
45 degrees flat bend, width						
150mm	nr	0.27	2.58	12.10	2.20	16.88
300mm	nr	0.33	3.16	12.36	2.33	17.85
450mm	nr	0.83	7.94	14.72	3.40	26.06
600mm	nr	1.00	9.57	24.71	5.14	39.42
750mm	nr	1.10	10.53	20.52	4.66	35.71
900mm	nr	1.20	11.48	23.17	5.20	39.85

Steel cable tray (cont'd)	Unit	Labour hours	Net labour (£)	Net material (£)	O'heads /profit (£)	Total (£)
equal tee, width						
150mm	nr	0.37	3.54	23.39	4.04	30.9?
300mm	nr	0.57	5.45	26.40	4.78	31.8?
450mm	nr	0.97	9.28	28.64	5.69	43.6?
600mm	nr	1.17	11.20	32.55	6.56	50.3?
750mm	nr	1.25	11.96	43.84	8.37	64.1?
900mm	nr	1.33	12.73	46.45	8.88	68.0?
four way, width						
150mm	nr	0.50	4.79	36.11	6.13	47.0?
300mm	nr	0.67	6.41	38.07	6.67	51.1?
450mm	nr	1.17	11.20	40.51	7.76	59.4?
600mm	nr	1.37	13.11	45.40	8.78	67.2?
750mm	nr	1.43	13.69	56.96	10.60	81.25
900mm	nr	1.50	14.36	60.01	11.16	85.5?
straight reducer, width						
300mm	nr	0.27	2.58	15.41	2.70	20.6?
450mm	nr	0.27	2.58	17.30	2.98	22.8?
600mm	nr	0.27	2.58	19.22	3.27	25.0?
750mm	nr	0.27	2.58	21.10	3.55	27.2?
900mm	nr	0.27	2.58	22.94	3.83	29.35
offset reducer, width						
300mm	nr	0.27	2.58	12.20	2.22	17.0?
450mm	nr	0.27	2.58	14.14	2.51	19.2?
600mm	nr	0.27	2.58	16.14	2.81	21.5?
750mm	nr	0.27	2.58	18.17	3.11	23.86
900mm	nr	0.27	2.58	20.01	3.39	25.98
couplers (pairs) all sizes	nr	0.54	5.17	2.94	1.22	9.33
Heavy duty, 130mm deep, gauge 2.0mm						
ladder 3m length, width						
150mm	m	0.33	3.16	10.87	2.10	16.13
300mm	m	0.57	5.45	11.78	2.58	19.81
450mm	m	0.57	5.45	12.84	2.74	21.03
600mm	m	0.80	7.66	13.73	3.21	24.60
750mm	nr	0.87	8.33	15.12	3.52	26.97
900mm	m	0.98	9.38	21.08	4.57	35.03

M & E MEASUREMENT SERVICES

	Unit	Labour hours	Net labour (£)	Net material (£)	O'heads /profit (£)	Total (£)
internal riser, width						
150mm	nr	0.33	3.16	26.89	4.51	34.56
300mm	nr	0.67	6.41	27.48	5.08	38.97
450mm	nr	0.83	7.94	29.79	5.66	43.39
600mm	nr	1.03	9.86	32.33	6.33	48.52
750mm	nr	1.10	10.53	35.62	6.92	53.07
900mm	nr	1.17	11.20	38.41	7.44	57.05
external riser, width						
150mm	nr	0.33	3.16	26.89	4.51	34.56
300mm	nr	0.67	6.41	27.48	4.33	33.89
450mm	nr	0.83	7.94	29.79	5.66	43.39
600mm	nr	1.03	9.86	32.33	6.33	48.52
750mm	nr	1.17	11.20	35.62	7.02	53.84
900mm	nr	1.17	11.20	38.41	7.44	57.05
90 degrees flat bend, width						
150mm	nr	0.27	2.58	18.37	3.14	24.09
300mm	nr	0.33	3.16	19.76	3.44	26.36
450mm	nr	0.83	7.94	22.99	4.64	35.57
600mm	nr	1.00	9.57	26.84	5.46	41.87
750mm	nr	1.10	10.53	30.30	6.12	46.95
900mm	nr	1.20	11.48	33.71	6.78	51.97
45 degrees flat bend, width						
150mm	nr	0.27	2.58	13.59	2.43	18.60
300mm	nr	0.33	3.16	14.20	2.60	19.96
450mm	nr	0.83	7.94	16.83	3.72	28.49
600mm	nr	1.00	9.57	20.44	4.50	34.51
750mm	nr	1.10	10.53	23.36	5.08	38.97
900mm	nr	1.20	11.48	26.38	5.68	43.54
equal tee, width						
150mm	nr	0.37	3.54	31.27	5.22	40.03
300mm	nr	0.57	5.45	35.41	6.13	46.99
450mm	nr	0.97	9.28	38.66	7.19	55.13
600mm	nr	1.17	11.20	46.93	8.72	66.85
750mm	nr	1.25	11.96	49.97	9.29	71.22
900mm	nr	1.33	12.73	53.31	9.91	75.95
four way, width						
150mm	nr	0.50	4.79	46.93	7.76	59.48
300mm	nr	0.60	5.74	49.97	8.36	64.07
450mm	nr	1.17	11.20	53.31	9.68	74.19

Steel cable tray (cont'd)	Unit	Labour hours	Net labour (£)	Net material (£)	O'heads /profit (£)	Total (£)
602mm	nr	1.37	13.11	59.79	10.93	83.83
752mm	nr	1.43	13.69	73.67	13.10	100.46
902mm	nr	1.50	14.36	77.47	13.77	105.60
straight reducer, width						
305mm	nr	0.27	2.58	18.69	3.19	24.46
452mm	nr	0.27	2.58	21.20	3.57	27.35
602mm	nr	0.27	2.58	23.57	3.92	30.07
752mm	nr	0.27	2.58	26.13	4.31	33.02
902mm	nr	0.27	2.58	28.70	4.69	35.97
offset reducer, width						
305mm	nr	0.27	2.58	15.32	2.69	17.90
452mm	nr	0.27	2.58	19.14	3.26	24.98
602mm	nr	0.27	2.58	20.13	3.41	26.12
752mm	nr	0.27	2.58	22.70	3.79	29.07
902mm	nr	0.27	2.58	25.27	4.18	32.03
couplers (pairs) all sizes	nr	0.54	5.17	5.11	1.54	11.82

Cable ladder system accessories

A discount of 10% has been incorporated within the material costs of these items

Twin cantilever bracket

ladder, width						
152mm	nr	0.43	4.12	0.88	0.75	5.75
305mm	nr	0.67	6.41	1.63	1.21	9.25
452mm	nr	0.93	8.90	2.99	1.78	13.67
602mm	nr	1.07	10.24	3.78	2.10	16.12
752mm	nr	1.12	10.72	5.98	2.50	19.20
902mm	nr	1.17	11.20	7.48	2.80	21.48

Cantilever bracket

ladder, width						
152mm	nr	0.43	4.12	0.73	0.73	5.58
305mm	nr	0.67	6.41	1.22	1.14	8.77
452mm	nr	0.93	8.90	1.93	1.62	12.45
602mm	nr	1.07	10.24	2.54	1.92	14.70
752mm	nr	1.12	10.72	4.63	2.30	17.65
902mm	nr	1.17	11.20	7.39	2.79	21.38

	Unit	Labour hours	Net labour (£)	Net material (£)	O'heads /profit (£)	Total (£)
Cantilever arm						
ladder, width						
150mm	nr	0.43	4.12	1.67	0.87	6.66
300mm	nr	0.67	6.41	2.28	1.30	9.99
450mm	nr	0.93	8.90	2.70	1.74	13.34
600mm	nr	2.07	19.81	3.29	3.46	26.56
750mm	nr	1.12	10.72	4.03	2.21	16.96
900mm	nr	1.17	11.20	4.94	2.42	18.56
Studding (3m length)						
thread size						
M6	m	0.12	1.15	0.55	0.26	1.96
M8	m	0.12	1.15	0.74	0.28	2.17
M10	m	0.12	1.15	0.78	0.29	2.22
M12	m	0.12	1.15	0.78	0.29	2.22
Hexagon head set screw						
thread size, length						
M6 x 20mm	100	0.30	2.87	2.73	0.84	6.44
M6 x 25mm	100	0.30	2.87	3.01	0.88	6.76
M6 x 30mm	100	0.30	2.87	5.82	1.30	9.99
M6 x 40mm	100	0.30	2.87	8.07	1.64	12.58
M8 x 20mm	100	0.30	2.87	4.77	1.15	8.79
M8 x 25mm	100	0.30	2.87	5.05	1.19	9.11
M8 x 35mm	100	0.30	2.87	6.02	1.33	10.22
M8 x 40mm	100	0.30	2.87	6.70	1.44	11.01
M10 x 25mm	100	0.30	2.87	6.90	1.47	11.24
M10 x 30mm	100	0.30	2.87	7.87	1.61	12.35
M10 x 35mm	100	0.30	2.87	7.87	1.61	12.35
M10 x 40mm	100	0.30	2.87	7.97	1.63	12.47
M12 x 20mm	100	0.30	2.87	10.39	1.99	15.25
M12 x 30mm	100	0.30	2.87	10.88	2.06	15.81
M12 x 35mm	100	0.30	2.87	12.14	2.25	17.26
M12 x 40mm	100	0.30	2.87	12.72	2.34	17.93

CONDUIT AND CABLE TRUNKING

Steel cable tray (cont'd)	Unit	Labour hours	Net labour (£)	Net material (£)	O'heads /profit (£)	Total (£)
Hexagon nut						
thread size						
M6	100	0.30	2.87	1.75	0.69	5.31
M8	100	0.30	2.87	1.85	0.71	5.43
M10	100	0.30	2.87	3.33	0.93	7.13
M12	100	0.30	2.87	6.81	1.45	11.13
Washers, plain						
thread, size						
M6	100	0.30	2.87	0.78	0.55	4.20
M8	100	0.30	2.87	1.06	0.59	4.52
M10	100	0.30	2.87	1.26	0.62	4.75
M12	100	0.30	2.87	1.75	0.69	5.31
Shakeproof						
thread, size						
M6	100	0.30	2.87	1.26	0.62	4.75
M8	100	0.30	2.87	1.55	0.66	5.08
M10	100	0.30	2.87	1.65	0.68	5.20
M12	100	0.30	2.87	2.14	0.75	5.76
Mushroom head bolt and nuts, M6						
thread, length						
12mm	100	0.33	3.16	3.10	0.94	7.20
16mm	100	0.33	3.16	4.66	1.17	8.99
20mm	100	0.33	3.16	5.05	1.23	9.44
25mm	100	0.33	3.16	5.55	1.31	10.02
30mm	100	0.33	3.16	6.12	1.39	10.67

	Unit	Labour hours	Net labour (£)	Net material (£)	O'heads /profit (£)	Total (£)
STEEL TRUNKING (WALSALL)						

General notes

1. All labour times include installing surface fixed trunking allowing for cutting, drilling of fixing holes and fitting cover once

2. All labour times and material costs include fixing trunking fittings to trunking and earth links measured in the running length

3. All labour times include installing materials up to a maximum height of 4.5m, but not erecting trestles or scaffolding. Add 10% to times for height 4.5-7m

4. Allow additional time for breaking into existing trunking installations

5. A discount of 36% has been incorporated within the nett material costs in this section

6. A waste factor of 5% has been incorporated in the running length of trunking

Galvanized steel trunking single compartment complete with lid with screw fixings, size

	Unit	Labour hours	Net labour (£)	Net material (£)	O'heads /profit (£)	Total (£)
50 x 50mm	m	0.30	2.87	2.86	0.86	6.59
75 x 50mm	m	0.32	3.06	3.40	0.97	7.43
75 x 75mm	m	0.37	3.54	3.69	1.08	8.31
100 x 50mm	m	0.37	3.54	4.00	1.13	8.67

CONDUIT AND CABLE TRUNKING

Steel trunking (cont'd)	Unit	Labour hours	Net labour (£)	Net material (£)	O'heads /profit (£)	Total (£)
100 x 75mm	m	0.42	4.02	4.90	1.34	10.26
100 x 100mm	m	0.48	4.59	5.08	1.45	11.12
150 x 50mm	m	0.55	5.26	5.79	1.66	12.71
150 x 75mm	m	0.55	5.26	6.50	1.76	13.52
150 x 100mm	m	0.57	5.45	7.11	1.88	14.44
150 x 150mm	m	0.58	5.55	8.39	2.09	16.03
200 x 100mm	m	0.66	6.32	8.87	2.28	17.47

Fittings; single compartment
fixed to trunking complete with
lids with screw fixings, size

	Unit	Labour hours	Net labour (£)	Net material (£)	O'heads /profit (£)	Total (£)
50 x 50mm	nr	0.10	0.96	3.10	0.61	4.67
75 x 50mm	nr	0.12	1.15	3.67	0.72	5.54
75 x 75mm	nr	0.12	1.15	3.93	0.76	5.84
100 x 50mm	nr	0.15	1.44	4.22	0.85	6.51
100 x 75mm	nr	0.15	1.44	4.51	0.89	6.84
100 x 100mm	nr	0.15	1.44	4.91	0.95	7.30
150 x 50mm	nr	0.20	1.91	5.59	1.12	7.50
150 x 75mm	nr	0.20	1.91	6.05	1.19	9.15
150 x 100mm	nr	0.20	1.91	6.32	1.23	9.46
150 x 150mm	nr	0.27	2.58	7.29	1.48	11.35
200 x 100mm	nr	0.30	2.87	12.48	2.30	17.65

90 degrees sharp bend, size

	Unit	Labour hours	Net labour (£)	Net material (£)	O'heads /profit (£)	Total (£)
50 x 50mm	nr	0.10	0.96	4.34	0.79	6.09
75 x 50mm	nr	0.12	1.15	4.83	0.90	6.88
75 x 75mm	nr	0.12	1.15	5.48	0.99	7.62
100 x 50mm	nr	0.15	1.44	5.96	1.11	8.51
100 x 75mm	nr	0.15	1.44	6.05	1.12	8.61
100 x 100mm	nr	0.15	1.44	6.51	1.19	9.14
150 x 50mm	nr	0.20	1.91	9.26	1.68	12.85
150 x 75mm	nr	0.20	1.91	10.33	1.84	14.08
150 x 100mm	nr	0.20	1.91	11.29	1.98	15.18
150 x 150mm	nr	0.27	2.58	11.86	2.17	16.61
220 x 100mm	nr	0.30	2.87	12.66	2.33	17.86

90 degrees gusset bend, size

	Unit	Labour hours	Net labour (£)	Net material (£)	O'heads /profit (£)	Total (£)
50 x 50mm	nr	0.10	0.96	2.86	0.57	4.39
75 x 50mm	nr	0.12	1.15	4.47	0.84	6.46
75 x 75mm	nr	0.12	1.15	3.60	0.71	5.46
100 x 50mm	nr	0.15	1.44	4.80	0.94	7.18

M & E MEASUREMENT SERVICES

	Unit	Labour hours	Net labour (£)	Net material (£)	O'heads /profit (£)	Total (£)
100 x 75mm	nr	0.15	1.44	4.43	0.88	6.75
100 x 100mm	nr	0.15	1.44	5.00	0.97	7.41
150 x 50mm	nr	0.20	1.91	5.48	1.11	8.50
150 x 75mm	nr	0.20	1.91	6.61	1.28	9.80
150 x 100mm	nr	0.20	1.91	6.67	1.29	9.87
150 x 150mm	nr	0.27	2.58	6.94	1.43	10.95
200 x 100mm	nr	0.30	2.87	12.13	2.25	17.25
Crossover, size						
50 x 50mm	nr	0.27	2.58	4.47	1.06	8.11
75 x 50mm	nr	0.34	3.25	4.80	1.21	9.26
75 x 75mm	nr	0.23	2.20	5.46	1.15	8.81
100 x 50mm	nr	0.37	3.54	6.07	1.44	11.05
100 x 75mm	nr	0.37	3.54	6.21	1.46	11.21
100 x 100mm	nr	0.37	3.54	6.53	1.51	11.58
150 x 50mm	nr	0.39	3.73	7.39	1.67	12.79
150 x 75mm	nr	0.41	3.92	7.93	1.78	13.63
150 x 100mm	nr	0.43	4.12	8.03	1.82	13.97
150 x 150mm	nr	0.44	4.21	10.07	2.14	16.42
200 x 100mm	nr	0.57	5.45	17.39	3.43	26.27
Reducers, size						
50 x 50 - 50 x 50mm	nr	0.25	2.39	5.78	1.23	9.40
75 x 75 - 50 x 50mm	nr	0.25	2.39	6.99	1.41	10.79
75 x 75 - 75 x 50mm	nr	0.25	2.39	6.99	1.41	10.79
100 x 50 - 50 x 50mm	nr	0.28	2.68	6.99	1.45	11.12
100 x 50 - 75 x 50mm	nr	0.28	2.68	6.99	1.45	11.12
100 x 75 - 50 x 50mm	nr	0.28	2.68	6.99	1.45	11.12
100 x 75 - 75 x 50mm	nr	0.28	2.68	6.99	1.45	11.12
100 x 75 - 75 x 75mm	nr	0.28	2.68	6.99	1.45	11.12
100 x 75 - 100 x 50mm	nr	0.28	2.68	6.99	1.45	11.12
100 x 100 - 50 x 50mm	nr	0.28	2.68	9.21	1.78	13.67
100 x 100 - 75 x 50mm	nr	0.28	2.68	9.21	1.78	13.67
100 x 100 - 75 x 75mm	nr	0.28	2.68	9.21	1.78	13.67
100 x 100 - 100 x 50mm	nr	0.28	2.68	9.21	1.78	13.67
100 x 100 - 100 x 75mm	nr	0.28	2.68	9.21	1.78	13.67
150 x 50 - 75 x 50mm	nr	0.30	2.87	9.21	1.81	13.89
150 x 75 - 75 x 50mm	nr	0.30	2.87	10.34	1.98	15.19
150 x 75 - 75 x 75mm	nr	0.30	2.87	10.34	2.88	13.21
150 x 75 - 100 x 75mm	nr	0.30	2.87	10.34	1.98	15.19
150 x 75 - 150 x 50mm	nr	0.30	2.87	10.34	1.98	15.19
150 x 100 - 50 x 50mm	nr	0.30	2.87	11.47	2.15	16.49

CONDUIT AND CABLE TRUNKING

Steel trunking (cont'd)	Unit	Labour hours	Net labour (£)	Net material (£)	O'heads /profit (£)	Total (£)
150 x 100 - 75 x 50mm	nr	0.30	2.87	11.47	2.15	16.49
150 x 100 - 75 x 75mm	nr	0.30	2.87	11.47	2.15	16.49
150 x 100 - 100 x 50mm	nr	0.30	2.87	11.47	2.15	16.49
150 x 100 - 100 x 75mm	nr	0.30	2.87	11.47	2.15	16.49
150 x 100 - 100 x 100mm	nr	0.30	2.87	11.47	2.15	16.49
150 x 100 - 150 x 50mm	nr	0.30	2.87	11.47	2.15	16.49
150 x 100 - 150 x 75mm	nr	0.30	2.87	11.47	2.15	16.49
150 x 150 - 75 x 75mm	nr	0.30	2.87	13.69	2.48	19.04
150 x 150 - 100 x 75mm	nr	0.32	3.06	13.69	2.51	19.26
150 x 150 - 100 x 100mm	nr	0.32	3.06	13.69	2.51	19.26
150 x 150 - 150 x 50mm	nr	0.32	3.06	13.69	2.51	19.26
150 x 150 - 150 x 75mm	nr	0.32	3.06	13.69	2.51	19.26
150 x 150 - 150 x 100mm	nr	0.32	3.06	13.69	2.51	19.26
200 x 100 - 150 x 100mm	nr	0.34	3.25	18.56	3.27	25.08
200 x 100 - 150 x 100mm	nr	0.34	3.25	18.56	3.27	25.08

Horizontal tee, size

50 x 50mm	nr	0.20	1.91	3.32	0.78	6.01
75 x 50mm	nr	0.25	2.39	3.76	0.92	7.07
75 x 75mm	nr	0.25	2.39	4.16	0.98	7.53
100 x 50mm	nr	0.27	2.58	4.31	1.03	7.92
100 x 75mm	nr	0.27	2.58	4.86	1.12	8.56
100 x 100mm	nr	0.27	2.58	5.16	1.16	8.90
150 x 50mm	nr	0.33	3.16	5.68	1.33	8.84
150 x 75mm	nr	0.33	3.16	5.82	1.35	10.33
150 x 100mm	nr	0.33	3.16	6.17	1.40	10.73
150 x 150mm	nr	0.33	3.16	8.27	1.71	13.14
200 x 100mm	nr	0.45	4.31	12.89	2.58	19.78

Universal tee, size

50 x 50mm	nr	0.20	1.91	4.47	0.96	7.34
75 x 50mm	nr	0.25	2.39	4.80	1.08	8.27
75 x 75mm	nr	0.25	2.39	5.62	1.20	9.21
100 x 50mm	nr	0.27	2.58	5.66	1.24	9.48
100 x 75mm	nr	0.27	2.58	6.57	1.37	10.52
100 x 100mm	nr	0.27	2.58	6.24	1.32	10.14
150 x 50mm	nr	0.33	3.16	6.62	1.47	11.25
150 x 75mm	nr	0.33	3.16	6.81	1.50	11.47
150 x 100mm	nr	0.33	3.16	6.73	1.48	11.37
150 x 150mm	nr	0.33	3.16	8.38	1.73	13.27
200 x 100mm	nr	0.45	4.31	13.25	2.63	20.19

	Unit	Labour hours	Net labour (£)	Net material (£)	O'heads /profit (£)	Total (£)
Accessories						
Trunking lids, fixed with screws, width						
50mm	nr	0.10	0.96	1.09	0.31	2.36
75mm	nr	0.10	0.96	1.62	0.39	2.97
100mm	nr	0.10	0.96	1.75	0.41	3.12
150mm	nr	0.10	0.96	2.55	0.53	4.04
255mm	nr	0.10	0.96	5.04	0.90	6.90
Straight connector, fixed with screws, trunking depth						
50mm	nr	0.17	1.63	0.91	0.38	2.92
75mm	nr	0.17	1.63	1.15	0.42	3.20
100mm	nr	0.17	1.63	1.35	0.43	2.98
150mm	nr	0.20	1.91	1.89	0.57	4.37
Flanged connector, fixed with screws, trunking depth						
50mm	nr	0.33	3.16	0.48	0.55	4.19
75mm	nr	0.42	4.02	0.82	0.73	5.57
100mm	nr	0.42	4.02	0.90	0.74	5.66
150mm	nr	0.58	5.55	1.13	1.00	7.68
Adaptor for joining fittings						
50 x 50mm	nr	0.17	1.63	2.68	0.65	4.96
75 x 50mm	nr	0.17	1.63	2.13	0.56	4.32
75 x 75mm	nr	0.17	1.63	3.45	0.76	5.84
100 x 50mm	nr	0.20	1.91	2.31	0.63	4.85
100 x 75mm	nr	0.20	1.91	2.16	0.61	4.68
100 x 100mm	nr	0.20	1.91	4.61	0.98	7.50
150 x 50mm	nr	0.25	2.39	3.26	0.85	6.50
150 x 75mm	nr	0.25	2.39	2.83	0.78	6.00
150 x 100mm	nr	0.25	2.39	2.91	0.80	6.10
150 x 150mm	nr	0.25	2.39	5.48	1.18	9.05
200 x 100mm	nr	0.33	3.16	10.69	2.08	15.93

CONDUIT AND CABLE TRUNKING

Steel trunking (cont'd)	Unit	Labour hours	Net labour (£)	Net material (£)	O'heads /profit (£)	Total (£)
Blank end, size						
50 x 50mm	nr	0.05	0.48	0.91	0.21	1.60
75 x 50mm	nr	0.05	0.48	0.97	0.22	1.67
75 x 75mm	nr	0.05	0.48	1.06	0.23	1.77
100 x 50mm	nr	0.05	0.48	1.06	0.23	1.77
100 x 75mm	nr	0.05	0.48	1.17	0.25	1.90
100 x 100mm	nr	0.05	0.48	1.42	0.28	2.18
150 x 50mm	nr	0.05	0.48	1.08	0.23	1.79
150 x 75mm	nr	0.05	0.48	1.20	0.25	1.93
150 x 100mm	nr	0.08	0.77	1.76	0.38	2.91
150 x 150mm	nr	0.08	0.77	2.09	0.43	3.29
200 x 100mm	nr	0.13	1.24	2.41	0.55	4.20
Fire barriers, fitted with trunking						
50 x 50mm	nr	0.33	3.16	3.85	1.05	8.06
75 x 50mm	nr	0.33	3.16	4.81	1.20	9.17
75 x 75mm	nr	0.33	3.16	5.78	1.34	10.28
100 x 50mm	nr	0.33	3.16	5.78	1.34	10.28
100 x 75mm	nr	0.33	3.16	6.74	1.49	11.39
100 x 100mm	nr	0.33	3.16	7.70	1.63	12.49
150 x 50mm	nr	0.33	3.16	7.70	1.63	12.49
150 x 75mm	nr	0.33	3.16	8.66	1.77	13.59
150 x 100mm	nr	0.33	3.16	9.63	1.92	14.71
150 x 150mm	nr	0.33	3.16	11.56	2.21	16.93
225 x 100mm	nr	0.33	3.16	11.55	2.21	16.92
Hanger bracket, fitted around trunking with screws						
Trunk size						
50 x 50mm	nr	0.20	1.91	1.20	0.47	3.58
75 x 75mm	nr	0.20	1.91	1.52	0.51	3.94
100 x 100mm	nr	0.20	1.91	1.86	0.57	4.34
150 x 75mm	nr	0.20	1.91	2.11	0.60	4.62
150 x 150mm	nr	0.20	1.91	2.79	0.70	5.40

M & E MEASUREMENT SERVICES

	Unit	Labour hours	Net labour (£)	Net material (£)	O'heads /profit (£)	Total (£)
Pin racks, size						
50 x 50mm	nr	0.17	1.63	3.60	0.78	6.01
75 x 50mm	nr	0.17	1.63	4.50	0.92	6.13
75 x 75mm	nr	0.17	1.63	5.40	1.05	8.08
100 x 50mm	nr	0.17	1.63	5.40	1.05	8.08
100 x 75mm	nr	0.17	1.63	6.29	1.19	9.11
100 x 100mm	nr	0.17	1.63	7.19	1.32	10.14
150 x 50mm	nr	0.17	1.63	7.19	1.32	10.14
150 x 75mm	nr	0.17	1.63	8.10	1.46	11.19
150 x 100mm	nr	0.17	1.63	9.00	1.59	12.22
150 x 150mm	nr	0.17	1.63	10.79	1.86	14.28
200 x 100mm	nr	0.17	1.63	10.79	1.86	14.28
Cable retaining strap, fitted to trunking, width						
50mm	nr	0.06	0.57	0.26	0.12	0.95
75mm	nr	0.06	0.57	0.28	0.13	0.98
100mm	nr	0.06	0.57	0.30	0.13	1.00
150mm	nr	0.06	0.57	0.32	0.13	1.02
200mm	nr	0.06	0.57	0.34	0.14	1.05
Cable trunking marking jig, size						
50mm	nr	0.00	0.00	0.71	0.11	0.82
75mm	nr	0.00	0.00	0.75	0.11	0.86
100mm	nr	0.00	0.00	0.91	0.14	1.05
150mm	nr	0.00	0.00	1.34	0.20	1.54
Earth continuity screws	100	0.00	0.00	2.96	0.44	3.40
Threaded cutting screw	100	0.00	0.00	3.52	0.53	4.05
Turnbuckle	100	0.00	0.00	3.17	0.48	3.17
M5 captive spring nut	100	0.00	0.00	4.83	0.72	4.83

CONDUIT AND CABLE TRUNKING

	Unit	Labour hours	Net labour (£)	Net material (£)	O'heads /profit (£)	Total (£)
Twin compartment cable trunking						
Galvanized steel trunking, twin compartment, complete with lid and connector, size						
50 x 50mm	m	0.30	2.87	3.77	1.00	6.64
75 x 50mm	m	0.32	3.06	4.32	1.11	8.49
75 x 75mm	m	0.37	3.54	4.95	1.27	9.76
100 x 50mm	m	0.37	3.54	4.91	1.27	9.72
100 x 75mm	m	0.42	4.02	6.16	1.53	11.71
100 x 100mm	m	0.48	4.59	6.56	1.67	12.82
150 x 50mm	m	0.55	5.26	6.70	1.72	11.96
150 x 75mm	m	0.55	5.26	7.76	1.95	14.97
150 x 100mm	m	0.57	5.45	8.59	2.11	16.15
150 x 150mm	m	0.58	5.55	10.53	2.41	18.49
200 x 100mm	m	0.66	6.32	13.24	2.93	22.49
Fixings, twin compartment fixed to trunking with screws, complete with lid						
90 degree sharp bend, size						
50 x 50mm	nr	0.10	0.96	5.23	0.93	7.12
75 x 50mm	nr	0.12	1.15	5.59	1.01	7.75
75 x 75mm	nr	0.12	1.15	6.55	1.16	8.86
100 x 50mm	nr	0.15	1.44	6.91	1.25	9.60
100 x 75mm	nr	0.15	1.44	6.54	1.20	9.18
100 x 100mm	nr	0.15	1.44	6.94	1.26	9.64
150 x 75mm	nr	0.20	1.91	10.45	1.85	14.21
150 x 100mm	nr	0.20	1.91	11.05	2.26	12.96
200 x 100mm	nr	0.27	2.58	12.50	2.55	15.08
90 degree gusset bend, size						
50 x 50mm	nr	0.10	0.96	3.81	0.72	5.49
75 x 50mm	nr	0.12	1.15	5.43	0.99	7.57
75 x 75mm	nr	0.12	1.15	4.69	0.88	6.72
100 x 50mm	nr	0.15	1.44	5.56	1.05	8.05
100 x 75mm	nr	0.15	1.44	5.30	1.01	7.75
100 x 100mm	nr	0.15	1.44	6.17	1.14	8.75
150 x 50mm	nr	0.20	1.91	6.61	1.28	9.80
150 x 75mm	nr	0.20	1.91	7.69	1.44	11.04

	Unit	Labour hours	Net labour (£)	Net material (£)	O'heads /profit (£)	Total (£)
150 x 100mm	nr	0.20	1.91	7.84	1.46	9.75
150 x 150mm	nr	0.27	2.58	8.13	1.61	12.32
200 x 100mm	nr	0.30	2.87	13.81	2.50	19.18
45 degree bend, size						
50 x 50mm	nr	0.10	0.96	3.80	0.71	5.47
75 x 50mm	nr	0.12	1.15	4.37	0.83	6.35
75 x 75mm	nr	0.12	1.15	4.81	0.89	6.85
100 x 50mm	nr	0.15	1.44	5.05	0.97	7.46
100 x 75mm	nr	0.15	1.44	5.58	1.05	8.07
100 x 100mm	nr	0.15	1.44	6.21	1.15	8.80
150 x 50mm	nr	0.20	1.91	6.93	1.33	10.17
150 x 75mm	nr	0.20	1.91	7.24	1.37	10.52
150 x 100mm	nr	0.20	1.91	7.35	1.39	10.65
150 x 150mm	nr	0.27	2.58	8.07	1.60	12.25
200 x 100mm	nr	0.30	2.87	13.67	2.48	19.02
Horizontal tee, size						
50 x 50mm	nr	0.20	1.91	4.55	0.97	7.43
75 x 50mm	nr	0.25	2.39	4.29	1.00	7.68
75 x 75mm	nr	0.25	2.39	5.65	1.21	9.25
100 x 50mm	nr	0.27	2.58	5.82	1.26	9.66
100 x 75mm	nr	0.27	2.58	6.43	1.35	10.36
100 x 100mm	nr	0.27	2.58	7.76	1.55	11.89
150 x 50mm	nr	0.33	3.16	9.36	1.88	14.40
150 x 75mm	nr	0.33	3.16	9.94	1.96	15.06
150 x 100mm	nr	0.33	3.16	10.28	2.02	15.46
150 x 150mm	nr	0.33	3.16	11.01	2.13	16.30
200 x 100mm	nr	0.45	4.31	15.99	3.04	23.34
Universal tee, size						
50 x 50mm	nr	0.20	1.91	5.84	1.16	8.91
75 x 50mm	nr	0.25	2.39	5.50	1.18	9.07
75 x 75mm	nr	0.25	2.39	7.13	1.43	10.95
100 x 50mm	nr	0.27	2.58	5.48	1.21	9.27
100 x 75mm	nr	0.27	2.58	7.23	1.47	11.28
100 x 100mm	nr	0.27	2.58	7.88	1.57	12.03
150 x 50mm	nr	0.33	3.16	8.59	1.76	13.51

CONDUIT AND CABLE TRUNKING

Steel trunking (cont'd)	Unit	Labour hours	Net labour (£)	Net material (£)	O'heads /profit (£)	Total (£)
150 x 75mm	nr	0.33	3.16	9.13	1.84	14.13
150 x 100mm	nr	0.33	3.16	9.69	1.93	14.78
150 x 150mm	nr	0.33	3.16	11.01	2.13	16.30
200 x 100mm	nr	0.33	3.16	14.74	2.69	20.59
Crossover, size						
50 x 50mm	nr	0.27	2.58	6.46	1.36	10.40
75 x 50mm	nr	0.33	3.16	6.09	1.39	10.64
75 x 75mm	nr	0.33	3.16	7.68	1.63	12.47
100 x 50mm	nr	0.37	3.54	5.87	1.41	10.82
100 x 75mm	nr	0.37	3.54	8.96	1.88	14.38
100 x 100mm	nr	0.37	3.54	11.42	2.24	17.20
150 x 50mm	nr	0.43	4.12	12.50	2.49	19.11
150 x 75mm	nr	0.43	4.12	13.45	2.64	20.21
150 x 100mm	nr	0.43	4.12	14.30	2.76	21.18
150 x 150mm	nr	0.43	4.12	15.59	2.96	22.67
200 x 100mm	nr	0.57	5.45	20.98	3.96	30.39
Reducers, size						
75 x 50mm	nr	0.25	2.39	7.18	1.44	11.01
75 x 75mm	nr	0.25	2.39	8.70	1.66	12.75
100 x 50mm	nr	0.28	2.68	8.70	1.71	13.09
100 x 75mm	nr	0.28	2.68	8.70	1.71	13.09
100 x 100mm	nr	0.28	2.68	11.47	2.12	16.27
150 x 50mm	nr	0.30	2.87	11.47	2.15	16.49
150 x 75mm	nr	0.30	2.87	12.88	2.36	18.11
150 x 100mm	nr	0.30	2.87	14.29	2.57	17.16
150 x 150mm	nr	0.32	3.06	17.06	3.02	23.14
200 x 100mm	nr	0.34	3.25	23.15	3.96	30.36

Triple compartment cable trunking

Galvanized trunking, steel triple compartment, complete with lid and connector, size

	Unit	Labour hours	Net labour (£)	Net material (£)	O'heads /profit (£)	Total (£)
75 x 50mm	m	0.32	3.06	5.24	1.24	9.54
75 x 75mm	m	0.37	3.54	6.21	1.46	11.21
100 x 50mm	m	0.37	3.54	5.83	1.41	10.78
100 x 75mm	m	0.42	4.02	7.42	1.72	13.16
100 x 100mm	m	0.48	4.59	8.04	1.89	14.52
150 x 50mm	m	0.55	5.26	7.62	1.93	14.81

	Unit	Labour hours	Net labour (£)	Net material (£)	O'heads /profit (£)	Total (£)
150 x 75mm	m	0.55	5.26	9.02	2.14	16.42
150 x 100mm	m	0.57	5.45	10.07	2.33	17.85
150 x 150mm	m	0.58	5.55	12.67	2.73	20.95
200 x 100mm	m	0.58	5.55	11.82	2.61	19.98

Fittings, triple compartment fixed to trunking with screws complete with lid

90 degrees sharp bend, size

75 x 50mm	nr	0.12	1.15	6.99	1.22	9.36
75 x 75mm	nr	0.12	1.15	7.94	1.36	10.45
100 x 50mm	nr	0.15	1.44	7.87	1.40	10.71
100 x 75mm	nr	0.15	1.44	6.98	1.26	9.68
100 x 100mm	nr	0.15	1.44	7.40	1.33	10.17
150 x 50mm	nr	0.20	1.91	9.45	1.70	13.06
150 x 75mm	nr	0.20	1.91	10.72	1.89	14.52
150 x 100mm	nr	0.20	1.91	12.42	2.15	16.48
150 x 150mm	nr	0.27	2.58	14.50	2.56	19.64
200 x 100mm	nr	0.30	2.87	16.00	2.83	21.70

90 degree gusset bend, size

75 x 50mm	nr	0.12	1.15	6.39	0.17	7.54
75 x 75mm	nr	0.12	1.15	5.77	0.17	6.92
100 x 50mm	nr	0.15	1.44	6.72	0.22	8.16
100 x 75mm	nr	0.15	1.44	6.60	0.22	8.04
100 x 100mm	nr	0.15	1.44	7.33	0.22	8.77
150 x 50mm	nr	0.20	1.91	7.89	0.29	9.80
150 x 75mm	nr	0.20	1.91	8.78	0.29	10.69
150 x 100mm	nr	0.20	1.91	9.01	0.29	10.92
150 x 150mm	nr	0.27	2.58	9.71	0.39	12.29
200 x 100mm	nr	0.30	2.87	14.98	0.43	17.85

45 degree bends, size

75 x 50mm	nr	0.12	1.15	5.28	0.96	7.39
75 x 75mm	nr	0.12	1.15	6.00	1.07	8.22
100 x 50mm	nr	0.15	1.44	5.89	1.10	8.43
100 x 75mm	nr	0.15	1.44	5.96	1.11	8.51
100 x 100mm	nr	0.15	0.00	6.63	0.99	7.62
150 x 50mm	nr	0.20	1.91	7.38	1.39	10.68
150 x 75mm	nr	0.20	1.91	7.68	1.44	11.03

CONDUIT AND CABLE TRUNKING

Steel trunking (cont'd)	Unit	Labour hours	Net labour (£)	Net material (£)	O'heads /profit (£)	Total (£)
150 x 100mm	nr	0.20	1.91	8.01	1.49	11.41
150 x 150mm	nr	0.27	2.58	9.04	1.74	13.36
200 x 100mm	nr	0.30	2.87	15.12	2.70	20.69
Horizontal tee, size						
75 x 50mm	nr	0.25	2.39	7.10	1.42	10.91
75 x 75mm	nr	0.27	2.58	7.87	1.57	12.02
100 x 50mm	nr	0.27	2.58	8.40	1.65	10.98
100 x 75mm	nr	0.27	2.58	8.98	1.73	13.29
100 x 100mm	nr	0.27	2.58	9.22	1.77	13.57
150 x 50mm	nr	0.33	3.16	9.74	1.94	14.84
150 x 75mm	nr	0.33	3.16	10.29	2.02	15.47
150 x 100mm	nr	0.33	3.16	10.59	2.06	15.81
150 x 150mm	nr	0.33	3.16	12.00	2.27	17.43
200 x 100mm	nr	0.35	3.35	17.09	3.07	23.51
Universal tee, size						
75 x 50mm	nr	0.25	2.39	7.99	1.56	11.94
75 x 75mm	nr	0.25	2.39	8.47	1.63	12.49
100 x 50mm	nr	0.27	2.58	8.99	1.74	13.31
100 x 75mm	nr	0.27	2.58	9.53	1.82	13.93
100 x 100mm	nr	0.27	2.58	10.12	1.90	14.60
150 x 50mm	nr	0.33	3.16	9.22	1.86	14.24
150 x 75mm	nr	0.33	3.16	9.49	1.90	14.55
150 x 100mm	nr	0.33	3.16	9.84	1.95	14.95
150 x 150mm	nr	0.33	3.16	11.22	2.16	16.54
200 x 100mm	nr	0.35	3.35	15.76	2.87	21.98
Crossover, size						
75 x 50mm	nr	0.34	3.25	10.07	2.00	15.32
75 x 75mm	nr	0.34	3.25	10.79	2.11	16.15
100 x 50mm	nr	0.37	3.54	11.01	2.18	16.73
100 x 75mm	nr	0.37	3.54	11.66	2.28	17.48
100 x 100mm	nr	0.37	3.54	12.46	2.40	18.40
150 x 50mm	nr	0.38	3.64	13.31	2.54	19.49
150 x 75mm	nr	0.41	3.92	14.19	2.72	20.83
150 x 100mm	nr	0.43	4.12	15.10	2.88	22.10
150 x 150mm	nr	0.44	4.21	16.34	3.08	23.63
200 x 100mm	nr	0.48	4.59	22.43	4.05	31.07

M & E MEASUREMENT SERVICES

	Unit	Labour hours	Net labour (£)	Net material (£)	O'heads /profit (£)	Total (£)
Reducers, size						
75 x 50mm	nr	0.25	2.39	7.13	1.43	10.95
75 x 75mm	nr	0.25	2.39	8.72	1.67	12.78
100 x 50mm	nr	0.28	2.68	8.70	1.71	13.09
100 x 75mm	nr	0.28	2.68	10.12	1.92	14.72
100 x 100mm	nr	0.30	2.87	11.50	2.16	16.53
150 x 50mm	nr	0.30	2.87	11.50	2.16	16.53
150 x 75mm	nr	0.30	2.87	12.92	2.37	18.16
150 x 100mm	nr	0.30	2.87	14.34	2.58	19.79
150 x 150mm	nr	0.32	3.06	17.13	3.03	23.22
200 x 100mm	nr	0.34	3.25	25.09	4.25	32.59

Sundry trunking and fittings

Bench trunking (supplied in 2m
lengths with lid and connector)

	Unit	Labour hours	Net labour (£)	Net material (£)	O'heads /profit (£)	Total (£)
single compartment trunking	m	0.32	3.06	8.65	1.76	11.71
90 degrees internal bend	nr	0.12	1.15	6.36	1.13	8.64
90 degrees external bend	nr	0.12	1.15	6.36	1.13	8.64
stop end	nr	0.05	0.48	1.02	0.22	1.72
153mm length lid cut for single socket	nr	0.10	0.96	1.09	0.31	2.36
228mm length lid cut for twin socket	nr	0.10	0.96	2.51	0.52	3.99
Telecom plate	nr	0.02	0.19	1.76	0.29	2.24
extra connector	nr	0.02	0.19	1.02	0.18	1.39

Skirting trunking (supplied in 2m
lengths with lid and connector)

	Unit	Labour hours	Net labour (£)	Net material (£)	O'heads /profit (£)	Total (£)
Twin compartment trunking	m	0.32	3.06	8.67	1.76	11.73
90 degrees internal bend	nr	0.12	1.15	6.31	1.12	8.58
90 degrees external bend	nr	0.12	1.15	6.32	1.12	8.59
90 degrees flat bend	nr	0.12	1.15	11.05	1.83	14.03
tee	nr	0.23	2.20	12.63	2.22	17.05
left hand stop end	nr	0.05	0.48	0.71	0.18	1.37
right hand stop end	nr	0.05	0.48	0.71	0.18	1.37
153mm length lid cut for single socket	nr	0.10	0.96	1.83	0.42	3.21
228mm length lid cut for twin socket	nr	0.10	0.96	2.45	0.51	3.92

197

CONDUIT AND CABLE TRUNKING

Steel trunking (cont'd)	Unit	Labour hours	Net labour (£)	Net material (£)	O'heads /profit (£)	Total (£)
Triple compartment trunking	m	0.32	3.06	11.26	2.15	14.32
90 degrees internal bend	nr	0.12	1.15	7.89	1.36	10.40
90 degrees external bend	nr	0.12	1.15	7.89	1.36	9.04
90 degrees flat bend	nr	0.12	1.15	13.81	2.24	17.20
tee	nr	0.23	2.20	15.78	2.70	20.68
left hand stop end	nr	0.05	0.48	0.87	0.20	1.55
right hand stop end	nr	0.05	0.48	0.87	0.20	1.55
153mm length lid cut for single socket	nr	0.10	0.96	2.21	0.48	3.65
228mm length lid cut for twin socket	nr	0.10	0.96	2.85	0.57	4.38
Sill trunking (supplied in 2m lengths with lid and connectors)						
single compartment trunking	m	0.32	3.06	6.99	1.51	10.05
90 degrees internal bend	nr	0.12	1.15	5.50	1.00	7.65
90 degrees external bend	nr	0.12	1.15	5.50	1.00	7.65
90 degrees flat bend	nr	0.12	1.15	1.56	0.41	3.12
tee	nr	0.23	2.20	10.99	1.98	15.17
left hand stop end	nr	0.05	0.48	0.68	0.17	1.33
right hand stop end	nr	0.05	0.48	0.68	0.17	1.33
153mm length lid cut for single socket	nr	0.10	0.96	1.67	0.39	3.02
228mm length lid cut for twin socket	nr	0.10	0.96	2.09	0.46	3.51
Twin compartment trunking	m	0.32	3.06	8.64	1.76	11.70
90 degrees internal bend	nr	0.12	1.15	6.30	1.12	8.57
90 degrees external bend	nr	0.12	1.15	6.30	1.12	8.57
90 degrees flat bend	nr	0.12	1.15	11.69	1.93	14.77
tee	nr	0.23	2.20	12.59	2.22	17.01
left hand stop end	nr	0.10	0.96	0.74	0.26	1.96
right hand stop end	nr	0.10	0.96	0.74	0.26	1.96
153mm length lid cut for single socket	nr	0.10	0.96	2.00	0.44	3.40
228mm length lid cut for twin socket	nr	0.10	0.96	2.59	0.53	4.08

M & E MEASUREMENT SERVICES

	Unit	Labour hours	Net labour (£)	Net material (£)	O'heads /profit (£)	Total (£)
Lighting trunking (supplied in 5m lengths)						
single compartment trunking	m	0.32	3.06	1.69	0.71	4.75
lid (PVC 2m long)	m	0.10	0.96	0.72	0.25	1.68
connector	nr	0.17	1.63	1.10	0.41	3.14
cable retainer	nr	0.02	0.19	0.17	0.05	0.41
hanger	nr	0.53	5.07	0.81	0.88	6.76
stop end	nr	0.05	0.48	0.43	0.14	1.05
fitting suspension	nr	0.30	2.87	0.50	0.51	3.88
bend	nr	0.12	1.15	2.90	0.61	4.66
tee	nr	0.23	2.20	3.45	0.85	6.50
crossover	nr	0.33	3.16	4.17	1.10	8.43
Underfloor trunking (supplied in 2m lengths complete with connectors)						
two equal compartments	m	0.33	3.16	8.16	1.70	11.32
horizontal bend	nr	0.33	3.16	12.41	2.34	17.91
riser bend	nr	0.33	3.16	7.81	1.65	12.62
stop end	nr	0.12	1.15	0.68	0.27	2.10
terminal box	nr	1.00	9.57	27.39	5.54	42.50
angle box	nr	1.17	11.20	27.39	5.79	44.38
through box	nr	1.17	11.20	27.39	5.79	44.38
tee box	nr	1.33	12.73	27.39	6.02	46.14
crossover box	nr	1.50	14.36	27.39	6.26	48.01
conduit box	nr	1.17	11.20	27.39	5.79	44.38
service outlet box	nr	1.25	11.96	36.07	7.20	55.23
three equal compartments	m	0.43	4.12	11.01	2.27	15.13
horizontal bend	nr	0.37	3.54	18.11	3.25	24.90
riser bend	nr	0.37	3.54	10.65	2.13	16.32
stop end	nr	0.12	1.15	0.85	0.30	2.30
terminal box	nr	1.00	9.57	28.29	5.68	43.54
angle box	nr	1.17	11.20	28.28	5.92	45.40
through box	nr	1.17	11.20	28.28	5.92	45.40
tee box	nr	1.33	12.73	28.28	6.15	47.16
crossover box	nr	1.50	14.36	28.28	6.40	49.04
conduit box	nr	1.17	11.20	28.28	5.92	45.40
service outlet box	nr	1.25	11.96	37.43	7.41	56.80

CONDUIT AND CABLE TRUNKING

Steel trunking (cont'd)	Unit	Labour hours	Net labour (£)	Net material (£)	O'heads /profit (£)	Total (£)
Flush floor shallow screed trunking (supplied in 2m lengths with lid and connector)						
two equal compartments	m	0.33	3.16	13.59	2.51	16.75
right angle bend	nr	0.33	3.16	13.35	2.48	18.99
Tee	nr	1.33	12.73	15.58	4.25	32.56
crossover	nr	1.50	14.36	15.58	4.49	29.94
riser bend	nr	0.33	3.16	15.58	2.81	21.55
stop end	nr	0.12	1.15	1.08	0.33	2.56
153mm length lid cut for single socket	nr	0.20	1.91	3.00	0.74	5.65
three equal compartments	m	0.43	4.12	19.74	3.58	23.86
right angle bend	nr	0.37	3.54	17.77	3.20	24.51
tee	nr	1.33	12.73	20.20	4.94	37.87
crossover	nr	1.50	14.36	20.20	5.18	39.74
riser bend	nr	0.37	3.54	17.77	3.20	24.51
stop end	nr	0.12	1.15	1.40	0.38	2.93
153mm length lid cut for single socket	nr	0.20	1.91	3.96	0.88	6.75
Flush floor trunking service outlet system (supplied in 1.5m lengths with lid and connectors)						
three compartment	m	0.43	4.12	26.35	4.57	30.47
riser bend	nr	0.37	3.54	21.49	3.75	28.78
stop end	nr	0.12	1.15	1.29	0.37	2.81
angle box complete with flyover	nr	1.17	11.20	26.72	5.69	43.61
tee box complete with flyover	nr	1.33	12.73	26.72	5.92	45.37
crossover box complete with flyover	nr	1.50	14.36	26.72	6.16	47.24
service outlet box without socket	nr	1.20	11.48	27.22	5.80	44.50
service outlet box with 2 unswitched single sockets	nr	1.25	11.96	31.66	6.54	43.62
service outlet box with a twin switched socket outlet	nr	1.25	11.96	31.66	6.54	50.16

M & E MEASUREMENT SERVICES

	Unit	Labour hours	Net labour (£)	Net material (£)	O'heads /profit (£)	Total (£)
Service outlet boxes for cavity floor installation						
service outlet box without socket	nr	1.25	11.96	34.39	6.95	53.30
service outlet box with 2 unswitched single sockets	nr	1.25	11.96	38.83	7.62	50.79
service outlet box with a twin switched socket outlet	nr	1.25	11.96	38.83	7.62	58.41

PVC compact mini trunking (white)

Mini trunking (MK EGA), size

	Unit	Labour hours	Net labour (£)	Net material (£)	O'heads /profit (£)	Total (£)
16 x 16mm	m	0.10	0.96	1.30	0.34	2.60
25 x 16mm	m	0.10	0.96	1.62	0.39	2.97
32 x 12.5mm	m	0.10	0.96	1.62	0.39	2.97
40 x 16mm	m	0.10	0.96	2.03	0.45	3.44
40 x 25mm	m	0.10	0.96	2.46	0.51	3.93
40 x 40mm	m	0.10	0.96	3.19	0.62	4.77
50 x 25mm	m	0.10	0.96	3.09	0.61	4.66
50 x 32mm	m	0.10	0.96	3.51	0.67	5.14

Twin compartment trunking, size

	Unit	Labour hours	Net labour (£)	Net material (£)	O'heads /profit (£)	Total (£)
40 x 16mm	m	0.05	0.48	2.39	0.43	3.30
40 x 25mm	m	0.05	0.48	2.91	0.51	3.90

Couplings, size

	Unit	Labour hours	Net labour (£)	Net material (£)	O'heads /profit (£)	Total (£)
16 x 16mm	nr	0.05	0.48	0.39	0.13	1.00
25 x 16mm	nr	0.05	0.48	0.39	0.13	1.00
32 x 12.5mm	nr	0.05	0.48	0.39	0.13	1.00
40 x 16mm	nr	0.05	0.48	0.39	0.13	1.00
40 x 25mm	nr	0.05	0.48	0.81	0.19	1.48
40 x 40mm	nr	0.05	0.48	1.39	0.28	2.15

Stop ends, size

	Unit	Labour hours	Net labour (£)	Net material (£)	O'heads /profit (£)	Total (£)
16 x 16mm	nr	0.05	0.48	0.39	0.13	1.00
25 x 16mm	nr	0.05	0.48	0.39	0.13	1.00
32 x 12.5mm	nr	0.05	0.48	0.39	0.13	1.00
40 x 16mm	nr	0.05	0.48	0.39	0.13	1.00
40 x 25mm	nr	0.05	0.48	0.48	0.14	1.10

CONDUIT AND CABLE TRUNKING

Steel trunking (cont'd)	Unit	Labour hours	Net labour (£)	Net material (£)	O'heads /profit (£)	Total (£)
40 x 40mm	nr	0.05	0.48	0.66	0.17	1.31
50 x 25mm	nr	0.08	0.77	0.71	0.22	1.70
50 x 32mm	nr	0.08	0.77	0.80	0.24	1.81
Flat angles, size						
16 x 16mm	nr	0.05	0.48	0.39	0.13	1.00
25 x 16mm	nr	0.05	0.48	0.39	0.13	1.00
32 x 12.5mm	nr	0.05	0.48	0.39	0.13	1.00
40 x 16mm	nr	0.05	0.48	0.39	0.13	1.00
40 x 25mm	nr	0.05	0.48	0.96	0.22	1.66
40 x 40mm	nr	0.05	0.48	1.34	0.27	2.09
50 x 25mm	nr	0.08	0.77	8.91	1.45	11.13
50 x 32mm	nr	0.08	0.77	9.74	1.58	12.09
Internal angles, size						
16 x 16mm	nr	0.05	0.48	0.39	0.13	1.00
25 x 16mm	nr	0.05	0.48	0.39	0.13	1.00
32 x 12.5mm	nr	0.05	0.48	0.39	0.13	1.00
40 x 16mm	nr	0.05	0.48	0.39	0.13	1.00
40 x 25mm	nr	0.05	0.48	0.96	0.22	1.66
40 x 40mm	nr	0.05	0.48	1.34	0.27	2.09
50 x 25mm	nr	0.08	0.77	8.91	1.45	11.13
50 x 32mm	nr	0.08	0.77	9.74	1.58	12.09
External angles, size						
16 x 16mm	nr	0.05	0.48	0.39	0.13	1.00
25 x 16mm	nr	0.05	0.48	0.39	0.13	1.00
32 x 12.5mm	nr	0.05	0.48	0.39	0.13	1.00
40 x 16mm	nr	0.05	0.48	0.39	0.13	1.00
40 x 25mm	nr	0.05	0.48	0.96	0.22	1.66
40 x 40mm	nr	0.05	0.48	1.34	0.27	2.09
50 x 25mm	nr	0.08	0.77	8.91	1.45	11.13
50 x 32mm	nr	0.08	0.77	9.74	1.58	12.09
Flat tees, size						
16 x 16mm	nr	0.05	0.48	0.66	0.17	1.31
25 x 16mm	nr	0.05	0.48	0.66	0.17	1.31
32 x 12.5mm	nr	0.05	0.48	0.66	0.17	1.31
40 x 16mm	nr	0.05	0.48	0.66	0.17	1.31
40 x 25mm	nr	0.05	0.48	0.96	0.22	1.66

M & E MEASUREMENT SERVICES

	Unit	Labour hours	Net labour (£)	Net material (£)	O'heads /profit (£)	Total (£)
40 x 40mm	nr	0.05	0.48	1.34	0.27	2.09
50 x 25mm	nr	0.08	0.77	7.76	1.28	9.81
50 x 32mm	nr	0.08	0.77	8.49	1.39	10.65
Adaptors, size						
16 x 16mm	nr	0.05	0.48	0.28	0.11	0.87
25 x 16mm	nr	0.05	0.48	0.28	0.11	0.87
32 x 12.5mm	nr	0.05	0.48	0.28	0.11	0.87
40 x 16mm	nr	0.05	0.48	0.28	0.11	0.87
Side mounting adaptors, size						
16 x 16mm	nr	0.05	0.48	0.85	0.20	1.53
25 x 16mm	nr	0.05	0.48	1.01	0.22	1.71
32 x 12.5mm	nr	0.05	0.48	1.01	0.22	1.71
40 x 16mm	nr	0.05	0.48	1.01	0.22	1.71
Side tees, size						
25 x 16/16 x 16mm	nr	0.05	0.48	0.70	0.18	1.36
25 x 16/25 x 16mm	nr	0.05	0.48	0.70	0.18	1.36
32 x 12.5mm	nr	0.05	0.48	0.70	0.18	1.36
Intersecting 3 way, size						
25 x 16mm RH	nr	0.05	0.48	0.66	0.17	1.31
25 x 16mm LH	nr	0.05	0.48	0.66	0.17	1.31
32 x 12.5mm RH	nr	0.05	0.48	0.66	0.17	1.31
32 x 12.5mm LH	nr	0.05	0.48	0.66	0.17	1.31
Circular boxes						
universal	nr	0.05	0.48	1.49	0.30	2.27
1 entry	nr	0.05	0.48	1.49	0.30	2.27
2 entry	nr	0.05	0.48	1.49	0.30	2.27

PVC communications trunking (white)

Trunking, standard, size

	Unit	Labour hours	Net labour (£)	Net material (£)	O'heads /profit (£)	Total (£)
11 x 8mm	m	0.13	1.24	0.72	0.29	2.25
16 x 10mm	m	0.13	1.24	0.82	0.31	2.37
20 x 12.5mm	m	0.13	1.24	0.91	0.32	2.47

CONDUIT AND CABLE TRUNKING

Steel trunking (cont'd)	Unit	Labour hours	Net labour (£)	Net material (£)	O'heads /profit (£)	Total (£)
Trunking 'speed fix', size						
11 x 8mm	m	0.08	0.77	1.10	0.28	2.15
16 x 10mm	m	0.08	0.77	1.43	0.33	2.53
20 x 12.5mm	m	0.08	0.77	1.97	0.41	3.15
Straight adaptor, size						
11 x 8mm	nr	0.05	0.48	0.26	0.11	0.85
16 x 10mm	nr	0.05	0.48	0.26	0.11	0.85
20 x 12.5mm	nr	0.05	0.48	0.26	0.11	0.85
Circular boxes						
1 entry	nr	0.05	0.48	1.49	0.30	2.27
2 entry	nr	0.05	0.48	1.49	0.30	2.27
Compact cornice trunking (white), supplied in 3m lengths						
40 x 40mm cornice trunking	m	0.17	1.63	3.99	0.84	6.46
end cap	nr	0.03	0.29	0.94	0.18	1.41
internal corner	nr	0.08	0.77	1.04	0.27	2.08
external corner	nr	0.08	0.77	1.04	0.27	2.08
intersection	nr	0.18	1.72	1.38	0.46	3.56
mini trunking adaptor	nr	0.08	0.77	1.41	0.33	2.18
cable retaining strap	nr	0.02	0.19	0.17	0.05	0.41
Standard cornice trunking (white), supplied in 3m lengths						
90 x 90mm cornice trunking	m	0.23	2.20	5.18	1.11	8.49
end cap	nr	0.03	0.29	1.13	0.21	1.63
internal corner	nr	0.08	0.77	1.25	0.30	2.32
external corner	nr	0.08	0.77	1.25	0.30	2.32
joint cover	nr	0.07	0.67	0.27	0.14	1.08
mini trunking adaptor	nr	0.08	0.77	1.71	0.37	2.48
cable retaining strap	nr	0.02	0.19	0.34	0.08	0.61
intersection LH	nr	0.20	1.91	1.80	0.56	4.27
intersection RH	nr	0.20	1.91	1.80	0.56	4.27

	Unit	Labour hours	Net labour (£)	Net material (£)	O'heads /profit (£)	Total (£)

PVC cable trunking (grey or white)

Trunking supplied in 3m lengths, size

50 x 50mm	m	0.17	1.63	5.92	1.13	8.68
75 x 50mm	m	0.17	1.63	6.80	1.26	9.69
75 x 75mm	m	0.20	1.91	8.44	1.55	11.90
100 x 50mm	m	0.20	1.91	9.71	1.74	13.36
100 x 75mm	m	0.22	2.11	10.78	1.93	14.82
100 x 100mm	m	0.23	2.20	12.47	2.20	16.87
150 x 75mm	m	0.28	2.68	21.07	3.56	27.31
150 x 100mm	m	0.30	2.87	26.40	4.39	33.66
150 x 150mm	m	0.33	3.16	40.10	6.49	49.75

Angle units, flat cover up to
75 x 75mm fabricated, no couplings
supplied, size

50 x 50mm	nr	0.08	0.77	3.53	0.64	4.94
75 x 50mm	nr	0.08	0.77	4.76	0.83	6.36
75 x 75mm	nr	0.08	0.77	5.86	0.99	7.62
100 x 50mm	nr	0.08	0.77	10.87	1.75	13.39
100 x 75mm	nr	0.08	0.77	15.60	2.46	18.83
100 x 100mm	nr	0.08	0.77	15.60	2.46	18.83
150 x 75mm	nr	0.10	0.96	25.70	4.00	30.66
150 x 100mm	nr	0.10	0.96	30.66	4.74	36.36
150 x 150mm	nr	0.10	0.96	44.77	6.86	52.59

Angle units, external cover,
50 x 50mm moulded, above 50 x 50mm
fabricated no couplings supplied,
size

50 x 50mm	nr	0.08	0.77	3.54	0.65	4.96
75 x 50mm	nr	0.08	0.77	7.93	1.30	10.00
75 x 75mm	nr	0.08	0.77	10.53	1.69	12.99
100 x 50mm	nr	0.08	0.77	11.89	1.90	14.56
100 x 75mm	nr	0.08	0.77	21.29	3.31	25.37
100 x 100mm	nr	0.08	0.77	21.29	3.31	25.37
150 x 75mm	nr	0.10	0.96	24.66	3.84	29.46
150 x 100mm	nr	0.10	0.96	29.31	4.54	34.81
150 x 150mm	nr	0.10	0.96	44.77	6.86	52.59

CONDUIT AND CABLE TRUNKING

PVC trunking (cont'd)	Unit	Labour hours	Net labour (£)	Net material (£)	O'heads /profit (£)	Total (£)
Angle units, internal cover, 50 x 50mm moulded, above 50 x 50mm fabricated, no couplings supplied, size						
50 x 50mm	nr	0.08	0.77	3.53	0.64	4.94
75 x 50mm	nr	0.08	0.77	7.93	1.30	10.00
75 x 75mm	nr	0.08	0.77	10.53	1.69	12.99
100 x 50mm	nr	0.08	0.77	11.89	1.90	14.56
100 x 75mm	nr	0.08	0.77	21.29	3.31	25.37
100 x 100mm	nr	0.08	0.77	21.29	3.31	25.37
150 x 75mm	nr	0.10	0.96	24.66	3.84	29.46
150 x 100mm	nr	0.10	0.96	29.31	4.54	34.81
150 x 150mm	nr	0.10	0.96	44.77	6.86	52.59
Tee units, flat cover, up to 75 x 75mm moulded, above 75 x 75mm fabricated, no couplings supplied, size						
50 x 50mm	nr	0.13	1.24	5.52	1.01	7.77
75 x 50mm	nr	0.13	1.24	7.02	1.24	9.50
75 x 75mm	nr	0.13	1.24	7.93	1.38	10.55
100 x 50mm	nr	0.15	1.44	12.84	2.14	16.42
100 x 75mm	nr	0.15	1.44	18.43	2.98	22.85
100 x 100mm	nr	0.15	1.44	18.43	2.98	22.85
150 x 75mm	nr	0.15	1.44	31.80	4.99	38.23
150 x 100mm	nr	0.15	1.44	40.80	6.34	48.58
150 x 150mm	nr	0.15	1.44	51.32	7.91	60.67
Tee units, external cover, fabricated, no couplings supplied, size						
50 x 50mm	nr	0.13	1.24	10.70	1.79	13.73
75 x 50mm	nr	0.13	1.24	16.80	2.71	20.75
75 x 75mm	nr	0.13	1.24	13.91	2.27	17.42
100 x 50mm	nr	0.15	1.44	16.59	2.70	20.73
100 x 75mm	nr	0.15	1.44	22.19	3.54	27.17
100 x 100mm	nr	0.15	1.44	22.19	3.54	27.17
150 x 75mm	nr	0.15	1.44	31.80	4.99	38.23
150 x 100mm	nr	0.15	1.44	40.80	6.34	48.58
150 x 150mm	nr	0.15	1.44	51.31	7.91	60.66

M & E MEASUREMENT SERVICES

	Unit	Labour hours	Net labour (£)	Net material (£)	O'heads /profit (£)	Total (£)
Tee units, internal cover fabricated, no couplings supplied, size						
50 x 50mm	nr	0.13	1.24	10.70	1.79	13.73
75 x 50mm	nr	0.13	1.24	11.68	1.94	14.86
75 x 75mm	nr	0.13	1.24	13.91	2.27	17.42
100 x 50mm	nr	0.15	1.44	16.59	2.70	20.73
100 x 75mm	nr	0.15	1.44	22.19	3.54	27.17
100 x 100mm	nr	0.15	1.44	22.19	3.54	27.17
150 x 75mm	nr	0.15	1.44	31.80	4.99	38.23
150 x 100mm	nr	0.15	1.44	40.80	6.34	48.58
150 x 150mm	nr	0.15	1.44	51.32	7.91	60.67
Reducers, fabricated, size						
75 x 50mm	nr	0.13	1.24	3.11	0.65	5.00
75 x 75mm	nr	0.13	1.24	4.34	0.84	6.42
100 x 50mm	nr	0.13	1.24	5.06	0.94	7.24
100 x 75mm	nr	0.13	1.24	6.19	1.11	8.54
100 x 100mm	nr	0.13	1.24	6.19	1.11	8.54
150 x 75mm	nr	0.13	1.24	10.29	1.73	13.26
150 x 100mm	nr	0.13	1.24	15.03	2.44	18.71
150 x 150mm	nr	0.13	1.24	15.03	2.44	18.71
Stop ends, moulded, size						
50 x 50mm	nr	0.03	0.29	0.73	0.15	1.17
75 x 50mm	nr	0.03	0.29	1.04	0.20	1.53
75 x 75mm	nr	0.03	0.29	1.38	0.25	1.92
100 x 50mm	nr	0.03	0.29	1.87	0.32	2.48
100 x 75mm	nr	0.03	0.29	2.91	0.48	3.68
100 x 100mm	nr	0.03	0.29	2.91	0.48	3.68
150 x 75mm	nr	0.05	0.48	7.71	1.23	9.42
150 x 100mm	nr	0.05	0.48	9.45	1.49	11.42
150 x 150mm	nr	0.05	0.48	13.14	2.04	15.66
Couplings, external with rivets, up to 100 x 100mm moulded, above 100 x 100mm fabricated, size						
50 x 50mm	nr	0.07	0.67	1.78	0.37	2.82
75 x 50mm	nr	0.07	0.67	2.45	0.47	3.59
75 x 75mm	nr	0.07	0.67	2.86	0.53	4.06

CONDUIT AND CABLE TRUNKING

PVC trunking (cont'd)	Unit	Labour hours	Net labour (£)	Net material (£)	O'heads /profit (£)	Total (£)
100 x 50mm	nr	0.07	0.67	4.54	0.78	5.99
100 x 75mm	nr	0.07	0.67	5.69	0.95	7.31
100 x 100mm	nr	0.07	0.67	5.69	0.95	7.31
150 x 75mm	nr	0.08	0.77	8.35	1.37	10.49
150 x 100mm	nr	0.08	0.77	10.31	1.66	12.74
150 x 150mm	nr	0.08	0.77	12.78	2.03	15.58
Couplings, internal plain, moulded, size						
50 x 50mm	nr	0.07	0.67	1.58	0.34	2.59
75 x 50mm	nr	0.07	0.67	1.74	0.36	2.77
75 x 75mm	nr	0.07	0.67	1.77	0.37	2.81
100 x 50mm	nr	0.07	0.67	2.36	0.45	3.48
100 x 75mm	nr	0.07	0.67	2.96	0.54	4.17
100 x 100mm	nr	0.07	0.67	2.96	0.54	4.17
Flanged coupling, fabricated, size						
50 x 50mm	nr	0.17	1.63	4.38	0.90	6.91
75 x 50mm	nr	0.22	2.11	5.00	1.07	8.18
75 x 75mm	nr	0.22	2.11	6.01	1.22	9.34
100 x 50mm	nr	0.30	2.87	6.55	1.41	10.83
100 x 75mm	nr	0.30	2.87	8.11	1.65	12.63
100 x 100mm	nr	0.30	2.87	8.11	1.65	12.63
150 x 75mm	nr	0.42	4.02	9.45	2.02	15.49
150 x 100mm	nr	0.42	4.02	11.75	2.37	18.14
150 x 150mm	nr	0.42	4.02	14.88	2.83	21.73
Bridge pieces						
BP/1 for CLT/1	nr	0.08	0.77	2.38	0.47	3.62
BP/2 for CLT/2, 3	nr	0.08	0.77	2.38	0.47	3.62
BP/3 for CLT/4, 5, 6	nr	0.10	0.96	4.11	0.76	5.83
Dividing fillets	nr	0.10	0.96	1.10	0.31	2.37

M & E MEASUREMENT SERVICES

	Unit	Labour hours	Net labour (£)	Net material (£)	O'heads /profit (£)	Total (£)
Dividing strip, 1.8m lengths						
VS/2 for CLT/1, 2, 4	nr	0.10	0.96	4.34	0.79	6.09
VS3 for CLT/3, 5, 7	nr	0.10	0.96	5.74	1.00	7.70
VS/4 for CLT/6, 8	nr	0.10	0.96	7.32	1.24	8.28
Pin racks	nr	0.17	1.63	9.91	1.73	13.27
Fluorescent light suspension units						
50mm	nr	0.53	5.07	1.25	0.95	7.27

	Unit	Labour hours	Net labour (£)	Net material (£)	O'heads /profit (£)	Total (£)

Y61 HV/LV CABLES AND WIRING

General notes

1. A discount of 70% has been incorporated within the nett material costs in this section excluding the MICC cable and accessories

2. A waste factor of 10% has been incorporated in all the cable length costs

Multicore armoured PVC insulated cables to BS6346 (copper/PVC/ SWA/PVC)

Clipped to surface -
2 cores - size mm2

	Unit	Labour hours	Net labour (£)	Net material (£)	O'heads /profit (£)	Total (£)
1.5	m	0.12	1.15	0.77	0.29	2.21
2.5	m	0.12	1.15	0.95	0.31	2.41
4	m	0.17	1.63	1.43	0.46	3.52
6	m	0.20	1.91	1.84	0.56	4.31
10	m	0.20	1.91	2.91	0.72	5.54
16	m	0.25	2.39	2.99	0.81	6.19

Clipped to tray, 2 cores, size mm2

	Unit	Labour hours	Net labour (£)	Net material (£)	O'heads /profit (£)	Total (£)
1.5	m	0.08	0.77	0.77	0.23	1.77
2.5	m	0.08	0.77	0.95	0.26	1.98
4	m	0.15	1.44	1.43	0.43	3.30
6	m	0.17	1.63	1.84	0.52	3.99
10	m	0.20	1.91	2.91	0.72	5.54
16	m	0.23	2.20	2.99	0.78	5.97

Laid in trenches or drawn into
ducts, 2 cores, size mm2

	Unit	Labour hours	Net labour (£)	Net material (£)	O'heads /profit (£)	Total (£)
1.5	m	0.03	0.29	0.77	0.16	1.22
2.5	m	0.03	0.29	0.95	0.19	1.43
4	m	0.05	0.48	1.43	0.29	2.20

M & E MEASUREMENT SERVICES

	Unit	Labour hours	Net labour (£)	Net material (£)	O'heads /profit (£)	Total (£)
6	m	0.05	0.48	1.84	0.35	2.67
10	m	0.08	0.77	2.91	0.55	4.23
16	m	0.08	0.77	2.99	0.56	4.32

Clipped to surface, 3 cores, size mm2

	Unit	Labour hours	Net labour (£)	Net material (£)	O'heads /profit (£)	Total (£)
1.5	m	0.12	1.15	0.93	0.31	2.39
2.5	m	0.12	1.15	1.14	0.34	2.63
4	m	0.17	1.63	1.81	0.52	3.96
6	m	0.20	1.91	2.42	0.65	4.98
10	m	0.20	1.91	3.87	0.87	6.65
16	m	0.25	2.39	4.75	1.07	8.21

Clipped to tray, 3 cores, size mm2

	Unit	Labour hours	Net labour (£)	Net material (£)	O'heads /profit (£)	Total (£)
1.5	m	0.08	0.77	0.93	0.26	1.96
2.5	m	0.08	0.77	1.14	0.29	2.20
4	m	0.15	1.44	1.81	0.49	3.74
6	m	0.17	1.63	2.42	0.61	4.66
10	m	0.20	1.91	3.87	0.87	6.65
16	m	0.23	2.20	4.75	1.04	7.99

Laid in trenches or drawn into ducts, 3 cores, size mm2

	Unit	Labour hours	Net labour (£)	Net material (£)	O'heads /profit (£)	Total (£)
1.5	m	0.05	0.48	0.93	0.21	1.62
2.5	m	0.05	0.48	1.14	0.24	1.86
4	m	0.07	0.67	1.81	0.37	2.85
6	m	0.07	0.67	2.42	0.46	3.55
10	m	0.10	0.96	3.87	0.72	5.55
16	m	0.10	0.96	4.75	0.86	6.57

Clipped to surface, 4 cores, size mm2

	Unit	Labour hours	Net labour (£)	Net material (£)	O'heads /profit (£)	Total (£)
1.5	m	0.12	1.15	1.00	0.32	2.47
2.5	m	0.12	1.15	1.27	0.36	2.78
4	m	0.17	1.63	2.29	0.59	4.51
6	m	0.20	1.91	2.78	0.70	5.39
10	m	0.20	1.91	3.97	0.88	6.76
16	m	0.30	2.87	4.85	1.16	8.88

Multicore PVC cables (cont'd)	Unit	Labour hours	Net labour (£)	Net material (£)	O'heads /profit (£)	Total (£)
Clipped to tray, 4 cores, size mm2						
1.5	m	0.08	0.77	1.00	0.27	2.04
2.5	m	0.08	0.77	1.27	0.31	2.35
4	m	0.15	1.44	2.29	0.56	4.29
6	m	0.17	1.63	2.78	0.66	5.07
10	m	0.20	1.91	3.97	0.88	6.76
16	m	0.23	2.20	4.85	1.06	8.11
Laid in trenches or drawn into ducts, 4 cores, size mm2						
1.5	m	0.07	0.67	1.00	0.25	1.92
2.5	m	0.07	0.67	1.27	0.29	2.23
4	m	0.08	0.77	2.29	0.46	3.52
6	m	0.08	0.77	2.78	0.53	4.08
10	m	0.12	1.15	3.97	0.77	5.89
16	m	0.12	1.15	4.85	0.90	6.90
Terminations for PVC insulated armoured cable including connection						
2 cores, size mm2						
1.5	nr	0.65	6.22	1.56	1.17	8.95
2.5	nr	0.73	6.99	1.56	1.28	9.83
4	nr	0.75	7.18	1.56	1.31	10.05
6	nr	0.80	7.66	1.92	1.44	11.02
10	nr	0.88	8.42	2.73	1.67	12.82
16	nr	0.95	9.09	2.73	1.77	13.59
3 cores, size mm2						
1.5	nr	0.65	6.22	1.56	1.17	8.95
2.5	nr	0.73	6.99	1.56	1.28	9.83
4	nr	0.75	7.18	1.56	1.31	10.05
6	nr	0.80	7.66	1.92	1.44	11.02
10	nr	0.88	8.42	2.73	1.67	12.82
16	nr	0.95	9.09	2.73	1.77	13.59

	Unit	Labour hours	Net labour (£)	Net material (£)	O'heads /profit (£)	Total (£)
4 cores, size mm2						
1.5	nr	0.65	6.22	1.56	1.17	8.95
2.5	nr	0.73	6.99	1.56	1.28	9.83
4	nr	0.75	7.18	1.56	1.31	10.05
6	nr	0.80	7.66	1.92	1.44	11.02
10	nr	0.88	8.42	2.73	1.67	12.82
16	nr	0.95	9.09	2.73	1.77	13.59

Single core cables having thermo-setting insulation and non-magnetic armour (copper/XPLE/SWA/PVC) to BS5467

Clipped to surface, 2 cores, size mm2

	Unit	Labour hours	Net labour (£)	Net material (£)	O'heads /profit (£)	Total (£)
1.5	m	0.12	1.15	0.76	0.29	2.20
2.5	m	0.12	1.15	0.94	0.31	2.40
4	m	0.17	1.63	1.41	0.46	3.50
6	m	0.20	1.91	1.80	0.56	4.27
10	m	0.20	1.91	2.93	0.73	5.57
16	m	0.25	2.39	2.95	0.80	6.14

Clipped to tray, 2 cores, size mm2

	Unit	Labour hours	Net labour (£)	Net material (£)	O'heads /profit (£)	Total (£)
1.5	m	0.08	0.77	0.76	0.23	1.76
2.5	m	0.08	0.77	0.94	0.26	1.97
4	m	0.15	1.44	1.41	0.43	3.28
6	m	0.17	1.63	1.80	0.51	3.94
10	m	0.20	1.91	2.93	0.73	5.57
16	m	0.23	2.20	2.95	0.77	5.92

Laid in trenches or drawn into 2 cores, size mm2

	Unit	Labour hours	Net labour (£)	Net material (£)	O'heads /profit (£)	Total (£)
1.5	m	0.03	0.29	0.76	0.16	1.21
2.5	m	0.03	0.29	0.94	0.18	1.41
4	m	0.05	0.48	1.41	0.28	2.17
6	m	0.05	0.48	1.80	0.34	2.62
10	m	0.08	0.77	2.93	0.56	4.26
16	m	0.08	0.77	2.95	0.56	4.28

Single core cables (cont'd)	Unit	Labour hours	Net labour (£)	Net material (£)	O'heads /profit (£)	Total (£)
Clipped to surface, 3 cores, size mm2						
1.5	m	0.12	1.15	0.92	0.31	2.38
2.5	m	0.12	1.15	1.13	0.34	2.62
4	m	0.17	1.63	1.79	0.51	3.93
6	m	0.20	1.91	2.40	0.65	4.96
10	m	0.20	1.91	3.93	0.88	6.72
16	m	0.25	2.39	4.90	1.09	8.38
Clipped to tray, 3 cores, size mm2						
1.5	m	0.08	0.77	0.92	0.25	1.94
2.5	m	0.08	0.77	1.13	0.28	2.18
4	m	0.15	1.44	1.79	0.48	3.71
6	m	0.17	1.63	2.40	0.60	4.63
10	m	0.20	1.91	3.93	0.88	6.72
16	m	0.23	2.20	4.90	1.06	8.16
Laid in trenches or drawn into 3 cores, size mm2						
1.5	m	0.05	0.48	0.92	0.21	1.61
2.5	m	0.05	0.48	1.13	0.24	1.85
4	m	0.07	0.67	1.79	0.37	2.83
6	m	0.07	0.67	2.40	0.46	3.53
10	m	0.10	0.96	3.93	0.73	5.62
16	m	0.10	0.96	4.90	0.88	6.74
Clipped to surface, 4 cores, size mm2						
1.5	m	0.12	1.15	1.03	0.33	2.51
2.5	m	0.12	1.15	1.31	0.37	2.83
4	m	0.17	1.63	2.33	0.59	4.55
6	m	0.20	1.91	2.84	0.71	5.46
10	m	0.20	1.91	4.00	0.89	6.80
16	m	0.30	2.87	4.93	1.17	8.97

M & E MEASUREMENT SERVICES

	Unit	Labour hours	Net labour (£)	Net material (£)	O'heads /profit (£)	Total (£)
Clipped to tray, 4 cores, size mm2						
1.5	m	0.08	0.77	1.03	0.27	2.07
2.5	m	0.08	0.77	1.31	0.31	2.39
4	m	0.15	1.44	2.33	0.57	4.34
6	m	0.17	1.63	2.84	0.67	5.14
10	m	0.20	1.91	4.00	0.89	6.80
16	m	0.23	2.20	4.93	1.07	8.20
Laid in trenches or drawn into 4 cores, size mm2						
1.5	m	0.07	0.67	1.03	0.26	1.96
2.5	m	0.07	0.67	1.31	0.30	2.28
4	m	0.08	0.77	2.33	0.46	3.56
6	m	0.08	0.77	2.84	0.54	4.15
10	m	0.12	1.15	4.00	0.77	5.92
16	m	0.12	1.15	4.93	0.91	6.99

Mineral insulated cables (MICC) fixed to backgrounds, including bends and straighting. Fixing measured elsewhere

A discount of 30% has been incorporated in the nett materials costs for MICC cables and accessories

Light duty cables, bare copper sheath

2 cores, size mm2

	Unit	Labour hours	Net labour (£)	Net material (£)	O'heads /profit (£)	Total (£)
1.0	m	0.08	0.77	1.00	0.27	2.04
1.5	m	0.08	0.77	1.20	0.30	2.27
2.5	m	0.10	0.96	1.56	0.38	2.90
4.0	m	0.12	1.15	2.31	0.52	3.98

3 cores, size mm2

	Unit	Labour hours	Net labour (£)	Net material (£)	O'heads /profit (£)	Total (£)
1.0	m	0.08	0.77	1.23	0.30	2.30
1.5	m	0.08	0.77	1.56	0.35	2.68
2.5	m	0.10	0.96	2.43	0.51	3.90

Mineral insulated cables (cont'd)	Unit	Labour hours	Net labour (£)	Net material (£)	O'heads /profit (£)	Total (£)
4 cores, size mm2						
1.0	m	0.08	0.77	1.47	0.34	2.58
1.5	m	0.08	0.77	1.89	0.40	3.06
2.5	m	0.10	0.96	2.94	0.58	4.48
7 cores, size mm2						
1.0	m	0.10	0.96	2.18	0.47	3.61
1.5	m	0.12	1.15	2.71	0.58	4.44
2.5	m	0.13	1.24	3.57	0.72	5.53
Light duty cables, LSF sheath						
2 cores, size mm2						
1.0	m	0.08	0.77	1.17	0.29	2.23
1.5	m	0.08	0.77	1.38	0.32	2.47
2.5	m	0.10	0.96	1.76	0.41	3.13
4.0	m	0.12	1.15	2.44	0.54	4.13
3 cores, size mm2						
1.0	m	0.08	0.77	1.42	0.33	2.52
1.5	m	0.08	0.77	1.74	0.38	2.89
2.5	m	0.10	0.96	2.43	0.51	3.90
4 cores, size mm2						
1.0	m	0.08	0.77	1.67	0.37	2.81
1.5	m	0.08	0.77	2.11	0.43	3.31
2.5	m	0.10	0.96	3.06	0.60	4.62
7 cores, size mm2						
1.0	m	0.10	0.96	2.51	0.52	3.99
1.5	m	0.12	1.15	3.10	0.64	4.89
2.5	m	0.13	1.24	3.97	0.78	5.99

M & E MEASUREMENT SERVICES

	Unit	Labour hours	Net labour (£)	Net material (£)	O'heads /profit (£)	Total (£)
Light duty cable, PVC sheath						
2 cores, size mm2						
1.0	m	0.08	0.77	1.17	0.29	2.23
1.5	m	0.08	0.77	1.32	0.31	2.40
2.5	m	0.10	0.96	1.69	0.40	3.05
4.0	m	0.12	1.15	2.44	0.54	4.13
3 cores, size mm2						
1.0	m	0.08	0.77	1.42	0.33	2.52
1.5	m	0.08	0.77	1.67	0.37	2.81
2.5	m	0.10	0.96	2.43	0.51	3.90
4 cores, size mm2						
1.0	m	0.08	0.77	1.67	0.37	2.81
1.5	m	0.08	0.77	2.03	0.42	3.22
2.5	m	0.10	0.96	3.06	0.60	4.62
7 cores, size mm2						
1.0	m	0.10	0.96	2.51	0.52	3.99
1.5	m	0.12	1.15	3.10	0.64	4.89
2.5	m	0.13	1.24	3.97	0.78	5.99
Heavy duty cables, bare copper						
1 core, size mm2						
10	m	0.08	0.77	2.24	0.45	3.46
16	m	0.12	1.15	3.07	0.63	4.85
25	m	0.13	1.24	4.30	0.83	6.37
35	m	0.15	1.44	5.72	1.07	8.23
50	m	0.17	1.63	7.28	1.34	8.91
70	m	0.20	1.91	9.46	1.71	13.08
95	m	0.22	2.11	12.42	2.18	16.71
120	m	0.23	2.20	15.18	2.61	19.99
150	m	0.30	2.87	18.83	3.25	24.95
185	m	0.38	3.64	22.92	3.98	30.54
240	m	0.47	4.50	29.76	5.14	39.40

Heavy duty cables (cont'd)	Unit	Labour hours	Net labour (£)	Net material (£)	O'heads /profit (£)	Total (£)
2 cores, size mm2						
1.5	m	0.08	0.77	1.90	0.40	3.07
2.5	m	0.10	0.96	2.34	0.49	3.79
4	m	0.12	1.15	2.95	0.61	4.71
6	m	0.15	1.44	3.93	0.81	6.18
10	m	0.17	1.63	5.10	1.01	7.74
16	m	0.20	1.91	7.35	1.39	10.65
25	m	0.23	2.20	10.32	1.88	14.40
3 cores, size mm2						
1.5	m	0.10	0.96	2.11	0.46	3.53
2.5	m	0.12	1.15	2.65	0.57	4.37
4	m	0.15	1.44	3.37	0.72	5.53
6	m	0.17	1.63	4.34	0.90	6.87
10	m	0.20	1.91	6.30	1.23	9.44
16	m	0.23	2.20	8.84	1.66	12.70
25	m	0.26	2.49	13.57	2.41	18.47
4 cores, size mm2						
1.5	m	0.12	1.15	2.63	0.57	4.35
2.5	m	0.15	1.44	3.30	0.71	5.45
4	m	0.17	1.63	4.13	0.86	6.62
6	m	0.20	1.91	5.51	1.11	8.53
10	m	0.23	2.20	7.82	1.50	11.52
16	m	0.26	2.49	11.42	2.09	16.00
25	m	0.29	2.78	16.59	2.91	22.28
7 cores, size mm2						
1.5	m	0.12	1.15	3.63	0.72	5.50
2.5	m	0.15	1.44	4.94	0.96	7.34
12 cores, size mm2						
2.5	m	0.17	1.63	8.62	1.54	11.79
19 cores, size mm2						
1.5	m	0.20	1.91	13.30	2.28	17.49

M & E MEASUREMENT SERVICES

	Unit	Labour hours	Net labour (£)	Net material (£)	O'heads /profit (£)	Total (£)

Heavy duty cables, LSF sheath

1 core, size mm2

10	m	0.08	0.77	2.44	0.48	3.69
16	m	0.12	1.15	3.33	0.67	5.15
25	m	0.13	1.24	4.61	0.88	6.73
35	m	0.15	1.44	6.05	1.12	8.61
50	m	0.17	1.63	7.65	1.39	10.67
70	m	0.20	1.91	9.94	1.78	13.63
95	m	0.22	2.11	13.04	2.27	17.42
120	m	0.23	2.20	15.92	2.72	20.84
150	m	0.30	2.87	19.62	3.37	25.86
185	m	0.38	3.64	24.07	4.16	31.87
240	m	0.47	4.50	31.04	5.33	40.87

2 cores, size mm2

1.5	m	0.08	0.77	2.15	0.44	3.36
2.5	m	0.10	0.96	2.55	0.53	4.04
4	m	0.12	1.15	3.23	0.66	5.04
6	m	0.15	1.44	4.27	0.86	6.57
10	m	0.17	1.63	5.51	1.07	8.21
16	m	0.20	1.91	7.78	1.45	11.14
25	m	0.23	2.20	11.03	1.98	15.21

3 cores, size mm2

1.5	m	0.10	0.96	2.37	0.50	3.83
2.5	m	0.12	1.15	2.92	0.61	4.68
4	m	0.15	1.44	3.71	0.77	5.92
6	m	0.17	1.63	4.67	0.94	7.24
10	m	0.20	1.91	6.72	1.29	9.92
16	m	0.23	2.20	9.49	1.75	13.44
25	m	0.26	2.49	14.31	2.52	19.32

4 cores, size mm2

1.5	m	0.12	1.15	2.86	0.60	4.61
2.5	m	0.15	1.44	3.60	0.76	5.80
4	m	0.17	1.63	4.47	0.91	7.01
6	m	0.20	1.91	5.87	1.17	8.95

HV/LV CABLES AND WIRING

Heavy duty cables (cont'd)	Unit	Labour hours	Net labour (£)	Net material (£)	O'heads /profit (£)	Total (£)
10	m	0.23	2.20	8.29	1.57	12.06
16	m	0.26	2.49	12.10	2.19	16.78
25	m	0.29	2.78	17.63	3.06	23.47
7 cores, size mm.						
1.5	m	0.12	1.15	3.97	0.77	5.89
2.5	m	0.15	1.44	5.35	1.02	7.81
12 cores, size mm2						
2.5	m	0.17	1.63	9.21	1.63	12.47
19 cores, size mm2						
1.5	m	0.20	1.91	14.02	2.39	18.32

Heavy duty cables, PVC sheath

1 core, size mm2

	Unit	Labour hours	Net labour (£)	Net material (£)	O'heads /profit (£)	Total (£)
10	m	0.08	0.77	2.44	0.48	3.69
16	m	0.12	1.15	3.33	0.67	5.15
25	m	0.13	1.24	4.61	0.88	6.73
35	m	0.15	1.44	6.05	1.12	8.61
50	m	0.17	1.63	7.65	1.39	10.67
70	m	0.20	1.91	9.94	1.78	13.63
95	m	0.22	2.11	13.04	2.27	17.42
120	m	0.23	2.20	15.92	2.72	20.84
150	m	0.30	2.87	19.62	3.37	25.86
185	m	0.38	3.64	24.07	4.16	31.87
240	m	0.47	4.50	31.04	5.33	40.87

2 cores, size mm2

	Unit	Labour hours	Net labour (£)	Net material (£)	O'heads /profit (£)	Total (£)
1.5	m	0.08	0.77	2.15	0.44	3.36
2.5	m	0.10	0.96	2.55	0.53	4.04
4	m	0.12	1.15	3.23	0.66	5.04
6	m	0.15	1.44	4.27	0.86	6.57
10	m	0.17	1.63	5.51	1.07	8.21
16	m	0.20	1.91	7.78	1.45	11.14
25	m	0.23	2.20	11.03	1.98	15.21

M & E MEASUREMENT SERVICES

	Unit	Labour hours	Net labour (£)	Net material (£)	O'heads /profit (£)	Total (£)
3 cores, size mm2						
1.5	m	0.10	0.96	2.37	0.50	3.83
2.5	m	0.12	1.15	2.92	0.61	4.68
4	m	0.15	1.44	3.71	0.77	5.92
6	m	0.17	1.63	4.67	0.94	7.24
10	m	0.20	1.91	6.72	1.29	9.92
16	m	0.23	2.20	9.49	1.75	13.44
25	m	0.26	2.49	14.31	2.52	19.32
4 cores, size mm2						
1.5	m	0.12	1.15	2.86	0.60	4.61
2.5	m	0.15	1.44	3.60	0.76	5.80
4	m	0.17	1.63	4.47	0.91	7.01
6	m	0.20	1.91	5.87	1.17	8.95
10	m	0.23	2.20	8.29	1.57	12.06
16	m	0.26	2.49	12.10	2.19	16.78
25	m	0.29	2.78	17.63	3.06	23.47
7 cores, size mm2						
1.5	m	0.12	1.15	3.97	0.77	5.89
2.5	m	0.15	1.44	5.33	1.02	7.79
12 cores, size mm2						
2.5	m	0.17	1.63	9.21	1.63	12.47
19 cores, size mm2						
1.5	m	0.20	1.91	14.02	2.39	18.32
Standard seal, fitted						
Plain, size mm						
20	nr	0.10	0.96	0.35	0.20	1.51
25	nr	0.10	0.96	0.97	0.29	2.22
32	nr	0.10	0.96	2.09	0.46	3.51
40	nr	0.10	0.96	3.90	0.73	5.59

Standard seals (cont'd)	Unit	Labour hours	Net labour (£)	Net material (£)	O'heads /profit (£)	Total (£)
Earth tail, size mm						
25	nr	0.10	0.96	2.35	0.50	3.81
32	nr	0.10	0.96	3.72	0.70	5.38
40	nr	0.10	0.96	5.60	0.98	7.54
Standard glands, fitted, size mm						
20	nr	0.10	0.96	0.85	0.27	2.08
25	nr	0.10	0.96	1.44	0.36	2.76
32	nr	0.10	0.96	2.80	0.56	4.32
40	nr	0.10	0.96	6.34	1.09	8.39

Space gland, 20mm to BS6081

	Unit	Labour hours	Net labour (£)	Net material (£)	O'heads /profit (£)	Total (£)
Brass locknuts, size mm						
20	nr	0.02	0.19	0.15	0.05	0.39
25	nr	0.02	0.19	0.33	0.08	0.60
32	nr	0.02	0.19	0.55	0.11	0.85
40	nr	0.02	0.19	1.40	0.24	1.83
Zinc plated steel lockwashers, size mm						
20	nr	0.02	0.19	0.06	0.04	0.29
25	nr	0.02	0.19	0.12	0.05	0.36
32	nr	0.02	0.19	0.15	0.05	0.39
40	nr	0.02	0.19	2.01	0.33	2.53
LSF gland shrouds, size mm						
20	nr	0.02	0.19	0.62	0.12	0.93
25	nr	0.02	0.19	0.93	0.17	1.29
32	nr	0.02	0.19	1.39	0.24	1.82
40	nr	0.02	0.19	2.23	0.36	2.78
PVC gland shrouds, size mm						
20	nr	0.02	0.19	0.23	0.06	0.48
25	nr	0.02	0.19	0.34	0.08	0.61
32	nr	0.02	0.19	0.41	0.09	0.69
40	nr	0.02	0.19	0.64	0.12	0.95

M & E MEASUREMENT SERVICES

	Unit	Labour hours	Net labour (£)	Net material (£)	O'heads /profit (£)	Total (£)
Light duty cable fixings, fixed to backgrounds, requiring drilling, plugging and screwing						
Bare copper, one hole clips						
2 to 7 core, all sizes	nr	0.18	1.72	0.04	0.26	2.02
PVC coated						
2 to 4 core, all sizes	nr	0.18	1.72	0.12	0.28	2.12
7 core	nr	0.18	1.72	0.16	0.28	2.16
Two-way saddles						
Bare copper						
2 to 7 core, all sizes	nr	0.22	2.11	0.04	0.32	2.47
PVC coated						
2 to 4 core, all sizes	nr	0.22	2.11	0.11	0.33	2.55
7 core	nr	0.22	2.11	0.14	0.34	2.59
Clips for multicore cables						
2 to 7 core, all sizes	nr	0.02	0.19	1.32	0.23	1.74
Heavy duty cable fixings, fixed to backgrounds requiring drilling, plugging and screwing						
Bare copper, one hole clips						
1 core, size mm2						
10	nr	0.10	0.96	0.04	0.15	1.15
16	nr	0.10	0.96	0.04	0.15	1.15
25	nr	0.10	0.96	0.05	0.15	1.16
35	nr	0.17	1.63	0.05	0.25	1.93
50	nr	0.17	1.63	0.10	0.26	1.99
70	nr	0.17	1.63	0.11	0.26	2.00
95	nr	0.17	1.63	0.11	0.26	2.00
120	nr	0.18	1.72	0.12	0.28	2.12

Cable fixings (cont'd)	Unit	Labour hours	Net labour (£)	Net material (£)	O'heads /profit (£)	Total (£)
150	nr	0.18	1.72	0.21	0.29	2.22
185	nr	0.18	1.72	0.23	0.29	2.24
240	nr	0.18	1.72	0.25	0.30	2.27
2 cores, size mm2						
1.5	nr	0.10	0.96	0.04	0.15	1.15
2.5	nr	0.10	0.96	0.05	0.15	1.16
4	nr	0.10	0.96	0.05	0.15	1.16
6	nr	0.17	1.63	0.09	0.26	1.98
10	nr	0.17	1.63	0.10	0.26	1.99
16	nr	0.17	1.63	0.11	0.26	2.00
25	nr	0.18	1.72	0.20	0.29	2.21
3 cores, size mm2						
1.5	nr	0.10	0.96	0.04	0.15	1.15
2.5	nr	0.10	0.96	0.05	0.15	1.16
4	nr	0.10	0.96	0.05	0.15	1.16
6	nr	0.17	1.63	0.09	0.26	1.98
10	nr	0.17	1.63	0.11	0.26	2.00
16	nr	0.17	1.63	0.11	0.26	2.00
25	nr	0.18	1.72	0.20	0.29	2.21
4 cores, size mm2						
1.5	nr	0.10	0.96	0.05	0.15	1.16
2.5	nr	0.10	0.96	0.05	0.15	1.16
4	nr	0.10	0.96	0.09	0.16	1.21
6	nr	0.17	1.63	0.10	0.26	1.99
10	nr	0.17	1.63	0.11	0.26	2.00
16	nr	0.17	1.63	0.20	0.27	2.10
25	nr	0.18	1.72	0.23	0.29	2.24
7 cores, size mm2						
1.5	nr	0.17	1.63	0.09	0.26	1.98
2.5	nr	0.17	1.63	0.10	0.26	1.99
12 cores, size mm2						
2.5	nr	0.17	1.63	0.11	0.26	2.00

M & E MEASUREMENT SERVICES

	Unit	Labour hours	Net labour (£)	Net material (£)	O'heads /profit (£)	Total (£)
19 cores, size mm2						
1.5	nr	0.18	1.72	0.12	0.28	2.12

Heavy duty cable, one hole clips, PVC coated

1 core, size mm2						
10	nr	0.10	0.96	0.12	0.16	1.24
16	nr	0.10	0.96	0.16	0.17	1.29
25	nr	0.10	0.96	0.18	0.17	1.31
35	nr	0.17	1.63	0.19	0.27	2.09
50	nr	0.17	1.63	0.20	0.27	2.10
70	nr	0.17	1.63	0.22	0.28	2.13
95	nr	0.17	1.63	0.31	0.29	2.23
120	nr	0.18	1.72	0.33	0.31	2.36
150	nr	0.18	1.72	0.35	0.31	2.38
185	nr	0.18	1.72	0.36	0.31	2.39
240	nr	0.18	1.72	0.40	0.32	2.44

2 cores, size mm2						
1.5	nr	0.10	0.96	0.12	0.16	1.24
2.5	nr	0.10	0.96	0.16	0.17	1.29
4	nr	0.10	0.96	0.18	0.17	1.14
6	nr	0.17	1.63	0.20	0.27	2.10
10	nr	0.17	1.63	0.20	0.27	2.10
16	nr	0.17	1.63	0.31	0.29	2.23
25	nr	0.18	1.72	0.33	0.31	2.36

3 cores, size mm2						
1.5	nr	0.10	0.96	0.16	0.17	1.29
2.5	nr	0.10	0.96	0.18	0.17	1.31
4	nr	0.10	0.96	0.19	0.17	1.15
6	nr	0.17	1.63	0.20	0.27	2.10
10	nr	0.17	1.63	0.22	0.28	2.13
16	nr	0.17	1.63	0.31	0.29	2.23
25	nr	0.18	1.72	0.35	0.31	2.38

Cable fixings (cont'd)	Unit	Labour hours	Net labour (£)	Net material (£)	O'heads /profit (£)	Total (£)
4 cores, size mm2						
1.5	nr	0.10	0.96	0.18	0.17	1.31
2.5	nr	0.10	0.96	0.19	0.17	1.32
4	nr	0.10	0.96	0.20	0.17	1.33
6	nr	0.17	1.63	0.20	0.27	2.10
10	nr	0.17	1.63	0.31	0.29	2.23
16	nr	0.17	1.63	0.33	0.29	2.25
25	nr	0.18	1.72	0.36	0.31	2.39
7 cores, size mm2						
1.5	nr	0.17	1.63	0.19	0.27	2.09
2.5	nr	0.17	1.63	0.12	0.26	2.01
12 cores, size mm2						
2.5	nr	0.17	1.63	0.31	0.29	2.23
19 cores, size mm2						
1.5	nr	0.18	1.72	0.32	0.31	2.35
Two-way saddles						
Bare copper, 1 core, size mm2						
10	nr	0.17	1.63	0.03	0.25	1.91
16	nr	0.17	1.63	0.04	0.25	1.92
25	nr	0.17	1.63	0.05	0.25	1.93
35	nr	0.20	1.91	0.05	0.29	2.25
50	nr	0.20	1.91	0.12	0.30	2.33
70	nr	0.20	1.91	0.12	0.30	2.33
95	nr	0.20	1.91	0.13	0.31	2.35
120	nr	0.22	2.11	0.21	0.35	2.67
150	nr	0.22	2.11	0.22	0.35	2.68
185	nr	0.22	2.11	0.23	0.35	2.69
240	nr	0.22	2.11	0.25	0.35	2.71
Bare copper, 2 cores, size mm2						
1.5	nr	0.17	1.63	0.04	0.25	1.92
2.5	nr	0.17	1.63	0.04	0.25	1.92
4	nr	0.17	1.63	0.05	0.25	1.93

M & E MEASUREMENT SERVICES

	Unit	Labour hours	Net labour (£)	Net material (£)	O'heads /profit (£)	Total (£)
6	nr	0.20	1.91	0.11	0.30	2.32
10	nr	0.20	1.91	0.12	0.30	2.33
16	nr	0.20	1.91	0.12	0.30	2.33
25	nr	0.22	2.11	0.21	0.35	2.67

Bare copper, 3 cores, size mm2

	Unit	Labour hours	Net labour (£)	Net material (£)	O'heads /profit (£)	Total (£)
1.5	nr	0.17	1.63	0.04	0.25	1.92
2.5	nr	0.17	1.63	0.05	0.25	1.93
4	nr	0.17	1.63	0.05	0.25	1.93
6	nr	0.20	1.91	0.11	0.30	2.32
10	nr	0.20	1.91	0.12	0.30	2.33
16	nr	0.20	1.91	0.13	0.31	2.35
25	nr	0.22	2.11	0.22	0.35	2.68

Bare copper, 4 cores, size mm2

	Unit	Labour hours	Net labour (£)	Net material (£)	O'heads /profit (£)	Total (£)
1.5	nr	0.17	1.63	0.05	0.25	1.93
2.5	nr	0.17	1.63	0.05	0.25	1.93
4	nr	0.17	1.63	0.11	0.26	2.00
6	nr	0.20	1.91	0.12	0.30	2.33
10	nr	0.20	1.91	0.12	0.30	2.33
16	nr	0.20	1.91	0.21	0.32	2.44
25	nr	0.22	2.11	0.23	0.35	2.69

Bare copper, 7 cores, size mm2

	Unit	Labour hours	Net labour (£)	Net material (£)	O'heads /profit (£)	Total (£)
1.5	nr	0.20	1.91	0.11	0.30	2.32
2.5	nr	0.20	1.91	0.12	0.30	2.33

Bare copper, 12 cores, size mm2

	Unit	Labour hours	Net labour (£)	Net material (£)	O'heads /profit (£)	Total (£)
2.5	nr	0.20	1.91	0.13	0.31	2.35

Bare copper, 19 cores, size mm2

	Unit	Labour hours	Net labour (£)	Net material (£)	O'heads /profit (£)	Total (£)
1.5	nr	0.22	2.11	0.21	0.35	2.67

PVC coated, 1 core, size mm2

	Unit	Labour hours	Net labour (£)	Net material (£)	O'heads /profit (£)	Total (£)
10	nr	0.17	1.63	0.11	0.26	2.00
16	nr	0.17	1.63	0.12	0.26	2.01
25	nr	0.17	1.63	0.20	0.27	2.10
35	nr	0.17	1.63	0.20	0.27	2.10

Cable fixings (cont'd)	Unit	Labour hours	Net labour (£)	Net material (£)	O'heads /profit (£)	Total (£)
50	nr	0.20	1.91	0.21	0.32	2.44
70	nr	0.20	1.91	0.24	0.32	2.47
95	nr	0.20	1.91	0.32	0.33	2.56
120	nr	0.22	2.11	0.33	0.37	2.81
150	nr	0.22	2.11	0.33	0.37	2.81
185	nr	0.22	2.11	0.41	0.38	2.90
240	nr	0.22	2.11	0.45	0.38	2.94
2 cores, size mm2						
0.5	nr	0.17	1.63	0.11	0.26	2.00
2.5	nr	0.17	1.63	0.12	0.26	2.01
4	nr	0.17	1.63	0.20	0.27	2.10
6	nr	0.20	1.91	0.20	0.32	2.43
10	nr	0.20	1.91	0.21	0.32	2.44
16	nr	0.20	1.91	0.32	0.33	2.56
25	nr	0.22	2.11	0.33	0.37	2.81
3 cores, size mm2						
1.5	nr	0.17	1.63	0.12	0.26	2.01
2.5	nr	0.17	1.63	0.20	0.27	2.10
4	nr	0.17	1.63	0.20	0.27	2.10
6	nr	0.20	1.91	0.21	0.32	2.44
10	nr	0.20	1.91	0.24	0.32	2.47
16	nr	0.20	1.91	0.33	0.34	2.58
25	nr	0.22	2.11	0.33	0.37	2.81
4 cores, size mm2						
1.5	nr	0.17	1.63	0.20	0.27	2.10
2.5	nr	0.17	1.63	0.20	0.27	2.10
4	nr	0.17	1.63	0.21	0.28	2.12
6	nr	0.20	1.91	0.21	0.32	2.44
10	nr	0.20	1.91	0.33	0.34	2.58
16	nr	0.20	1.91	0.33	0.34	2.58
25	nr	0.22	2.11	0.41	0.38	2.90
7 cores, size mm2						
1.5	nr	0.20	1.91	0.20	0.32	2.43
2.5	nr	0.20	1.91	0.21	0.32	2.44

	Unit	Labour hours	Net labour (£)	Net material (£)	O'heads /profit (£)	Total (£)
12 cores, size mm2						
2.5	nr	0.20	1.91	0.33	0.34	2.58
19 cores, size mm2						
1.5	nr	0.22	2.11	0.33	0.37	2.81
Clamp for single core cable						
1 core, size mm2						
10	nr	0.10	0.96	0.05	0.15	1.16
16	nr	0.10	0.96	0.05	0.15	1.16
25	nr	0.10	0.96	0.05	0.15	1.16
35	nr	0.17	1.63	0.07	0.25	1.95
50	nr	0.17	1.63	0.07	0.25	1.95
70	nr	0.17	1.63	0.07	0.25	1.95
95	nr	0.17	1.63	0.07	0.25	1.95
120	nr	0.18	1.72	0.09	0.27	2.08
150	nr	0.18	1.72	0.09	0.27	2.08
185	nr	0.18	1.72	0.11	0.27	2.10
240	nr	0.18	1.72	0.11	0.27	2.10
Clips for multicore cables						
2 cores, size mm2						
1.5	nr	0.10	0.96	0.03	0.15	1.14
2.5	nr	0.10	0.96	0.03	0.15	1.14
4	nr	0.10	0.96	0.03	0.15	1.14
6	nr	0.17	1.63	0.03	0.25	1.91
10	nr	0.17	1.63	0.03	0.25	1.91
16	nr	0.17	1.63	0.03	0.25	1.91
25	nr	0.18	1.72	0.04	0.26	2.02
3 cores, size mm2						
1.5	nr	0.10	0.96	0.03	0.15	1.14
2.5	nr	0.10	0.96	0.03	0.15	1.14
4	nr	0.10	0.96	0.03	0.15	1.14
6	nr	0.17	1.63	0.03	0.25	1.91

Cable fixings (cont'd)	Unit	Labour hours	Net labour (£)	Net material (£)	O'heads /profit (£)	Total (£)
10	nr	0.17	1.63	0.03	0.25	1.91
16	nr	0.17	1.63	0.03	0.25	1.91
25	nr	0.18	1.72	0.04	0.26	2.02
4 cores, size mm2						
1.5	nr	0.10	0.96	0.03	0.15	1.14
2.5	nr	0.10	0.96	0.03	0.15	1.14
4	nr	0.10	0.96	0.03	0.15	1.14
6	nr	0.17	1.63	0.03	0.25	1.91
10	nr	0.17	1.63	0.03	0.25	1.91
16	nr	0.17	1.63	0.04	0.25	1.92
25	nr	0.18	1.72	0.04	0.26	2.02
7 cores, size mm2						
1.5	nr	0.17	1.63	0.03	0.25	1.91
2.5	nr	0.17	1.63	0.03	0.25	1.91
12 cores, size mm2						
2.5	nr	0.17	1.63	0.03	0.25	1.91
19 cores, size mm2						
1.5	nr	0.18	1.72	0.04	0.26	2.02

Conductor insulating sleeving, maximum operating temperature 105 degrees centigrade, PVC sleeving or PVC extension sleeving (per 100mm)

All items are measured per 100mm and have units of dm

Conductor, size mm2

	Unit	Labour hours	Net labour (£)	Net material (£)	O'heads /profit (£)	Total (£)
1 (per 100mm)	dm	0.03	0.29	0.03	0.05	0.32
1.5 (per 100mm)	dm	0.03	0.29	0.03	0.05	0.32
2.5 (per 100mm)	dm	0.03	0.29	0.03	0.05	0.32
4 (per 100mm)	dm	0.03	0.29	0.03	0.05	0.32
6 (per 100mm)	dm	0.03	0.29	0.03	0.05	0.32
10 (per 100mm)	dm	0.03	0.29	0.04	0.05	0.33

	Unit	Labour hours	Net labour (£)	Net material (£)	O'heads /profit (£)	Total (£)
16 (per 100mm)	dm	0.04	0.38	0.04	0.06	0.42
25 (per 100mm)	dm	0.04	0.38	0.04	0.06	0.42
35 (per 100mm)	dm	0.04	0.38	0.05	0.06	0.43
50 (per 100mm)	dm	0.05	0.48	0.05	0.08	0.53
70 (per 100mm)	dm	0.05	0.48	0.05	0.08	0.53
95 (per 100mm)	dm	0.06	0.57	0.06	0.09	0.63
120 (per 100mm)	dm	0.08	0.77	0.06	0.12	0.83
150 (per 100mm)	dm	0.08	0.77	0.08	0.13	0.85
185 (per 100mm)	dm	0.09	0.86	0.09	0.14	0.95
240 (per 100mm)	dm	0.10	0.96	0.10	0.16	1.06

**Maximum operating temperature
150 degrees centigrade, silicon
elastomer coated glass sleeving
(per 100mm)**

Conductor size mm2

1 (per 100mm)	dm	0.03	0.29	0.07	0.05	0.36
1.5 (per 100mm)	dm	0.03	0.29	0.07	0.05	0.36
2.5 (per 100mm)	dm	0.03	0.29	0.07	0.05	0.36
4 (per 100mm)	dm	0.03	0.29	0.08	0.06	0.37
6 (per 100mm)	dm	0.03	0.29	0.08	0.06	0.37
10 (per 100mm)	dm	0.03	0.29	0.09	0.06	0.38
16 (per 100mm)	dm	0.04	0.38	0.11	0.07	0.49
25 (per 100mm)	dm	0.04	0.38	0.13	0.08	0.51
35 (per 100mm)	dm	0.04	0.38	0.16	0.08	0.54
50 (per 100mm)	dm	0.05	0.48	0.16	0.10	0.64
70 (per 100mm)	dm	0.05	0.48	0.19	0.10	0.67
95 (per 100mm)	dm	0.06	0.57	0.23	0.12	0.80
120 (per 100mm)	dm	0.08	0.77	0.23	0.15	1.00
150 (per 100mm)	dm	0.08	0.77	0.39	0.17	1.16
185 (per 100mm)	dm	0.09	0.86	0.39	0.19	1.25
240 (per 100mm)	dm	0.10	0.96	0.42	0.21	1.38

**Maximum operating temperature
250 degrees centigrade 100mm
single headed PTFE**

Conductor size mm2

1 (per 100mm)	dm	0.03	0.29	0.59	0.13	0.88
1.5 (per 100mm)	dm	0.03	0.29	0.59	0.13	0.88
2.5 (per 100mm)	dm	0.03	0.29	0.59	0.13	0.88

Sleeving (cont'd)	Unit	Labour hours	Net labour (£)	Net material (£)	O'heads /profit (£)	Total (£)
4 (per 100mm)	dm	0.03	0.29	0.60	0.13	0.89
6 (per 100mm)	dm	0.03	0.29	0.60	0.13	0.89
10 (per 100mm)	dm	0.03	0.29	0.64	0.14	0.93
16 (per 100mm)	dm	0.04	0.38	0.69	0.16	1.07
25 (per 100mm)	dm	0.04	0.38	0.79	0.18	1.17
35 (per 100mm)	dm	0.04	0.38	0.84	0.18	1.22
50 (per 100mm)	dm	0.05	0.48	0.84	0.20	1.32
70 (per 100mm)	dm	0.05	0.48	1.01	0.22	1.49
95 (per 100mm)	dm	0.06	0.57	1.22	0.27	1.79
120 (per 100mm)	dm	0.08	0.77	1.49	0.34	2.26
150 (per 100mm)	dm	0.08	0.77	1.91	0.40	2.68
185 (per 100mm)	dm	0.09	0.86	2.01	0.43	2.87
240 (per 100mm)	dm	0.10	0.96	2.22	0.48	3.18

Conductor connectors joining ferrules for straight through joints

Conductor, size mm2

	Unit	Labour hours	Net labour (£)	Net material (£)	O'heads /profit (£)	Total (£)
1	nr	0.10	0.96	0.10	0.16	1.22
1.5	nr	0.10	0.96	0.12	0.16	1.24
2.5	nr	0.10	0.96	0.12	0.16	1.24
4	nr	0.10	0.96	0.21	0.18	1.35
6	nr	0.10	0.96	0.36	0.20	1.52
10	nr	0.10	0.96	0.37	0.20	1.53
16	nr	0.12	1.15	0.50	0.25	1.90
25	nr	0.12	1.15	0.57	0.26	1.98
35	nr	0.12	1.15	0.59	0.26	2.00
50	nr	0.15	1.44	0.62	0.31	2.37
120	nr	0.20	1.91	0.99	0.43	3.33
150	nr	0.22	2.11	1.40	0.53	4.04
185	nr	0.25	2.39	2.59	0.75	5.73
240	nr	0.28	2.68	3.05	0.86	6.59

Cable terminal lugs compression type

Conductor, size mm2

	Unit	Labour hours	Net labour (£)	Net material (£)	O'heads /profit (£)	Total (£)
1	nr	0.08	0.77	0.09	0.13	0.99
1.5	nr	0.08	0.77	0.11	0.13	1.01
2.5	nr	0.08	0.77	0.11	0.13	1.01
4	nr	0.08	0.77	0.17	0.14	1.08

	Unit	Labour hours	Net labour (£)	Net material (£)	O'heads /profit (£)	Total (£)
6	nr	0.08	0.77	0.17	0.14	1.08
10	nr	0.08	0.77	0.26	0.15	1.18
16	nr	0.09	0.86	0.56	0.21	1.63
25	nr	0.09	0.86	0.61	0.22	1.69
35	nr	0.10	0.96	1.09	0.31	2.36
50	nr	0.12	1.15	1.38	0.38	2.91
70	nr	0.12	1.15	1.68	0.42	3.25
95	nr	0.12	1.15	2.22	0.51	3.88
120	nr	0.14	1.34	2.89	0.63	4.86
150	nr	0.15	1.44	2.89	0.65	4.98
185	nr	0.17	1.63	3.85	0.82	6.30
240	nr	0.20	1.91	5.46	1.11	8.48

Fire resistant cable 'FP200' range - Pirelli

Clipped to surfaces
No of cores   Size mm2

		Unit	Labour hours	Net labour (£)	Net material (£)	O'heads /profit (£)	Total (£)
2	1.0	m	0.12	1.15	1.05	0.33	2.53
2	1.5	m	0.12	1.15	1.27	0.36	2.78
2	2.5	m	0.13	1.24	1.62	0.43	3.29
3	1.0	m	0.12	1.15	1.33	0.37	2.85
3	1.5	m	0.12	1.15	1.66	0.42	3.23
3	2.5	m	0.13	1.24	2.06	0.49	3.79
4	1.0	m	0.13	1.24	1.62	0.43	3.29
4	1.5	m	0.13	1.24	2.02	0.49	3.75

Clipped to tray
No of Cores   Size mm2

		Unit	Labour hours	Net labour (£)	Net material (£)	O'heads /profit (£)	Total (£)
2	1.0	m	0.03	0.29	1.05	0.20	1.54
2	1.5	m	0.03	0.29	1.27	0.23	1.79
2	2.5	m	0.05	0.48	1.62	0.32	2.42
3	1.0	m	0.03	0.29	1.33	0.24	1.86
3	1.5	m	0.03	0.29	1.66	0.29	2.24
3	2.5	m	0.05	0.48	2.06	0.38	2.92
4	1.0	m	0.03	0.29	1.62	0.29	2.20
4	1.5	m	0.03	0.29	2.02	0.35	2.66

Fire resistant cable (cont'd)	Unit	Labour hours	Net labour (£)	Net material (£)	O'heads /profit (£)	Total (£)

Single core PVC insulated PVC sheathed cable BS6004 (supplied in packs of 50m)

Clipped to surface, size mm2

1	m	0.07	0.67	0.16	0.12	0.95
1.5	m	0.07	0.67	0.21	0.13	1.01
2.5	m	0.07	0.67	0.35	0.15	1.17
4	m	0.07	0.67	0.57	0.19	1.43
6	m	0.08	0.77	0.76	0.23	1.76
10	m	0.08	0.77	1.22	0.30	2.29
16	m	0.08	0.77	1.59	0.35	2.71
25	m	0.10	0.96	3.00	0.45	3.96
35	m	0.11	1.05	4.60	0.85	6.50

Fixed in chases covered with galvanized or PVC sheath, size mm2

1	m	0.10	0.96	0.16	0.17	1.29
1.5	m	0.10	0.96	0.21	0.18	1.35
2.5	m	0.10	0.96	0.35	0.20	1.51
4	m	0.10	0.96	0.57	0.23	1.76
6	m	0.12	1.15	0.76	0.29	2.20
10	m	0.12	1.15	1.22	0.36	2.73
16	m	1.59	15.22	1.59	2.52	19.33
25	m	0.14	1.34	3.00	0.65	4.99
35	m	0.15	1.44	4.60	0.91	6.95

Single core PVC insulated cables, non-armoured with sheath (twin and earth) BS6004 (supplied in packs of 50m)

Clipped to surface, size mm2

1	m	0.07	0.67	0.27	0.14	1.08
1.5	m	0.07	0.67	0.35	0.15	1.17
2.5	m	0.07	0.67	0.49	0.17	1.33
4	m	0.07	0.67	1.03	0.26	1.96
6	m	0.08	0.77	1.40	0.33	2.50
10	m	0.08	0.77	2.50	0.49	3.76
16	m	0.08	0.77	3.89	0.70	5.36

	Unit	Labour hours	Net labour (£)	Net material (£)	O'heads /profit (£)	Total (£)
Fixed in chases, covered with galvanized or PVC sheath, size mm2						
1	m	0.10	0.96	0.27	0.18	1.41
1.5	m	0.10	0.96	0.35	0.20	1.51
2.5	m	0.10	0.96	0.49	0.22	1.67
4	m	0.10	0.96	1.03	0.30	2.29
6	m	0.12	1.15	1.40	0.38	2.93
10	m	0.12	1.15	2.50	0.55	4.20
16	m	0.12	1.15	3.89	0.76	5.80

Single core PVC insulated cable, non-armoured non-sheathed (6491X singles)

Drawn in conduit, size mm2

1	m	0.02	0.19	0.07	0.04	0.30
1.5	m	0.02	0.19	0.11	0.04	0.34
2.5	m	0.03	0.29	0.17	0.07	0.53
4	m	0.03	0.29	0.28	0.09	0.66
6	m	0.04	0.38	0.40	0.12	0.90
10	m	0.04	0.38	0.82	0.18	1.38

Installed in trunking, size mm2

1	m	0.02	0.19	0.07	0.04	0.30
1.5	m	0.02	0.19	0.11	0.04	0.34
2.5	m	0.02	0.19	0.17	0.05	0.41
4	m	0.02	0.19	0.28	0.07	0.54
6	m	0.03	0.29	0.40	0.10	0.79
10	m	0.03	0.29	0.82	0.17	1.28
16	m	0.03	0.29	1.22	0.23	1.74
25	m	0.04	0.38	1.89	0.34	2.61
35	m	0.04	0.38	2.50	0.43	3.31
50	m	0.05	0.48	3.65	0.62	4.75
70	m	0.05	0.48	5.01	0.82	6.31
95	m	0.06	0.57	7.73	1.25	9.55
120	m	0.07	0.67	10.63	1.70	13.00
150	m	0.08	0.77	13.28	2.11	16.16

HV/LV CABLES AND WIRING

	Unit	Labour hours	Net labour (£)	Net material (£)	O'heads /profit (£)	Total (£)
Single core PVC insulated PVC sheathed cable with integral earth wire to BS6004						
Clipped to surface, size mm2						
1	m	0.07	0.67	0.23	0.14	1.04
1.5	m	0.07	0.67	0.30	0.15	1.12
Fixed in chases covered with galvanized or PVC sheath, size mm2						
1	m	0.07	0.67	0.23	0.14	1.04
1.5	m	0.07	0.67	0.30	0.15	1.12

M & E MEASUREMENT SERVICES

	Unit	Labour hours	Net labour (£)	Net material (£)	O'heads /profit (£)	Total (£)

Y70 HV SWITCHGEAR

General notes

1. All labour times allow for fixing equipment to backgrounds including any drilling, plugging and screwing

2. All labour times on switches and protection equipment allow for making standard cable connections

3. No discounts have been incorporated to any of the prices in this section due to the variance of discounts on equipment

Air circuit breakers tested to BS4572 (Ottermill Ltd) Automatic 50kA 500V P2 ratings, 1P31/1P54, vent/non-vent board

3-pole, manual/spring, fixed frame, size

	Unit	Labour hours	Net labour (£)	Net material (£)	O'heads /profit (£)	Total (£)
OM8 800/800A	nr	6.00	57.42	1331.00	208.26	1596.68
OM8 1000/1000A	nr	6.00	57.42	1363.00	213.06	1633.48
OM8 1200/1200A	nr	6.00	57.42	1373.00	214.56	1644.98
OMA10 1000/1000A	nr	6.00	57.42	1649.00	255.96	1962.38
OMA10 1250/1250A	nr	6.00	57.42	1673.00	259.56	1730.42
OMA10 1600/1250A	nr	6.00	57.42	1698.00	263.31	1755.42
OMA12 2000/2000A	nr	6.00	57.42	2129.00	327.96	2514.38
OMA12 2500/2000A	nr	6.00	57.42	2275.00	349.86	2682.28
OMA16 3000/2500A	nr	6.00	57.42	2805.00	429.36	3291.78
OMA16 3200/2500A	nr	6.00	57.42	3416.00	521.01	3473.42

Air circuit breakers (cont'd)	Unit	Labour hours	Net labour (£)	Net material (£)	O'heads /profit (£)	Total (£)
3-pole, manual/spring, draw out cassetted, frame size						
OM8 800/800A	nr	6.00	57.42	1793.00	277.56	2127.98
OM8 1000/1000A	nr	6.00	57.42	1813.00	280.56	2150.98
OM8 1200/1200A	nr	6.00	57.42	1844.00	285.21	2186.63
OMA10 1000/1000A	nr	6.00	57.42	2172.00	334.41	2563.83
OMA10 1250/1250A	nr	6.00	57.42	2205.00	339.36	2601.78
OMA10 1600/1250A	nr	6.00	57.42	2247.00	345.66	2650.08
OMA12 2000/2000A	nr	6.00	57.42	2767.00	423.66	3248.08
OMA12 2500/2000A	nr	6.00	57.42	3087.00	471.66	3616.08
OMA16 3000/2500A	nr	6.00	57.42	3622.00	551.91	4231.33
OMA16 3200/2500A	nr	6.00	57.42	3886.00	591.51	4534.93
OMA24 3400/3000A	nr	6.00	57.42	5637.00	854.16	6548.58
OMA24 4000/3000A	nr	6.00	57.42	6679.00	1010.46	7746.88
3-pole, solenoid 110V DC, draw out cassetted, frame size						
OM8 800/800A	nr	6.00	57.42	2174.00	334.71	2566.13
OM8 1000/1000A	nr	6.00	57.42	2195.00	337.86	2590.28
OM8 1200/1200A	nr	6.00	57.42	2224.00	342.21	2623.63
OMA10 1000/1000A	nr	6.00	57.42	2553.00	391.56	3001.98
OMA10 1250/1250A	nr	6.00	57.42	2587.00	396.66	3041.08
OMA10 1600/1250A	nr	6.00	57.42	2620.00	401.61	3079.03
OMA12 2000/2000A	nr	6.00	57.42	3146.00	480.51	3683.93
OMA12 2500/2000A	nr	6.00	57.42	3467.00	528.66	4053.08
OMA16 3000/2500A	nr	6.00	57.42	4005.00	609.36	4671.78
OMA16 3200/2500A	nr	6.00	57.42	4296.00	653.01	5006.43
OMA24 3400/3000A	nr	6.00	57.42	6091.00	922.26	7070.68
4-pole, manual/spring, draw out cassetted, frame size						
OM8 800/800A	nr	6.00	57.42	2226.00	342.51	2283.42
OM8 1000/1000A	nr	6.00	57.42	2259.00	347.46	2663.88
OM8 1200/1200A	nr	6.00	57.42	2306.00	354.51	2717.93
OMA10 1000/1000A	nr	6.00	57.42	2707.00	414.66	3179.08
OMA10 1250/1250A	nr	6.00	57.42	2742.00	419.91	3219.33
OMA10 1600/1250A	nr	6.00	57.42	2770.00	424.11	3251.53
OMA12 2000/2000A	nr	6.00	57.42	3385.00	516.36	3958.78
OMA12 2500/2000A	nr	6.00	57.42	3992.00	607.41	4656.83
OMA16 3000/2500A	nr	6.00	57.42	4737.00	719.16	5513.58
OMA16 3200/2500A	nr	6.00	57.42	5093.00	772.56	5922.98

	Unit	Labour hours	Net labour (£)	Net material (£)	O'heads /profit (£)	Total (£)
4-pole, solenoid 110V DC, draw out cassetted, frame size						
OM8 800/800A	nr	6.00	57.42	2605.00	399.36	3061.78
OM8 1000/1000A	nr	6.00	57.42	2721.00	416.76	3195.18
OM8 1200/1200A	nr	6.00	57.42	2721.00	416.76	3195.18
OMA10 1000/1000A	nr	6.00	57.42	3087.00	471.66	3616.08
OMA10 1250/1250A	nr	6.00	57.42	3121.00	476.76	3655.18
OMA10 1600/1250A	nr	6.00	57.42	3149.00	480.96	3687.38
OMA12 2000/2000A	nr	6.00	57.42	3767.00	573.66	4398.08
OMA12 2500/2000A	nr	6.00	57.42	4453.00	676.56	5186.98
OMA16 3000/2500A	nr	6.00	57.42	5280.00	800.61	6138.03
OMA16 3200/2500A	nr	6.00	57.42	5649.00	855.96	6562.38
Non-automatic 50kA 500V P2 ratings, IP31/IP54						
3-pole, manual/spring, fixed frame, size						
OM8 800/800A	nr	6.00	57.42	1168.00	183.81	1409.23
OM8 1000/1000A	nr	6.00	57.42	1189.00	186.96	1433.38
OM8 1200/1200A	nr	6.00	57.42	1211.00	190.26	1458.68
OMA10 1000/1000A	nr	6.00	57.42	1352.00	211.41	1620.83
OMA10 1250/1250A	nr	6.00	57.42	1399.00	218.46	1456.42
OMA10 1600/1250A	nr	6.00	57.42	1452.00	226.41	1735.83
OMA12 2000/2000A	nr	6.00	57.42	1846.00	285.51	2188.93
OMA12 2500/2000A	nr	6.00	57.42	1944.00	300.21	2301.63
OMA16 3000/2500A	nr	6.00	57.42	2311.00	355.26	2723.68
OMA16 3200/2500A	nr	6.00	57.42	2505.00	384.36	2946.78
3-pole, manual/spring, draw out cassetted, frame size						
OM8 800/800A	nr	6.00	57.42	1620.00	251.61	1929.03
OM8 1000/1000A	nr	6.00	57.42	1642.00	254.91	1699.42
OM8 1200/1200A	nr	6.00	57.42	1678.00	260.31	1995.73
OMA10 1000/1000A	nr	6.00	57.42	1843.00	285.06	2185.48
OMA10 1250/1250A	nr	6.00	57.42	1875.00	289.86	2222.28
OMA10 1600/1250A	nr	6.00	57.42	1907.00	294.66	2259.08
OMA12 2000/2000A	nr	6.00	57.42	2438.00	374.31	2869.73
OMA12 2500/2000A	nr	6.00	57.42	2757.00	422.16	3236.58

Air circuit breakers (cont'd)	Unit	Labour hours	Net labour (£)	Net material (£)	O'heads /profit (£)	Total (£)
OMA16 3000/2500A	nr	6.00	57.42	3127.00	477.66	3662.08
OMA16 3200/2500A	nr	6.00	57.42	3429.00	522.96	4009.38
OMA24 3400/3000A	nr	6.00	57.42	5061.00	767.76	5886.18
OMA24 4000/3000A	nr	6.00	57.42	6104.00	924.21	7085.63

3-pole, solenoid 110V DC, draw
out cassetted, frame size

OM8 800/800A	nr	6.00	57.42	2000.00	308.61	2366.03
OM8 1000/1000A	nr	6.00	57.42	2023.00	312.06	2392.48
OM8 1200/1200A	nr	6.00	57.42	2057.00	317.16	2431.58
OMA10 1000/1000A	nr	6.00	57.42	2219.00	341.46	2617.88
OMA10 1250/1250A	nr	6.00	57.42	2255.00	346.86	2659.28
OMA10 1600/1250A	nr	6.00	57.42	2287.00	351.66	2696.08
OMA12 2000/2000A	nr	6.00	57.42	2818.00	431.31	3306.73
OMA12 2500/2000A	nr	6.00	57.42	3139.00	479.46	3675.88
OMA16 3000/2500A	nr	6.00	57.42	3511.00	535.26	4103.68
OMA16 3200/2500A	nr	6.00	57.42	3801.00	578.76	4437.18
OMA24 3400/3000A	nr	6.00	57.42	5516.00	836.01	6409.43
OMA24 4000/3000A	nr	6.00	57.42	6558.00	992.31	7607.73

4-pole, manual/spring, draw out
cassetted, frame size

OM8 800/800A	nr	6.00	57.42	2032.00	313.41	2402.83
OM8 1000/1000A	nr	6.00	57.42	2060.00	317.61	2435.03
OM8 1200/1200A	nr	6.00	57.42	2141.00	329.76	2528.18
OMA10 1000/1000A	nr	6.00	57.42	2261.00	347.76	2666.18
OMA10 1250/1250A	nr	6.00	57.42	2825.00	0.00	2882.42
OMA10 1250/1250A	nr	6.00	57.42	2444.00	375.21	2876.63
OMA10 1600/1250A	nr	6.00	57.42	2506.00	384.51	2947.93
OMA12 2000/2000A	nr	6.00	57.42	3055.00	466.86	3579.28
OMA12 2500/2000A	nr	6.00	57.42	3663.00	558.06	4278.48
OMA16 3000/2500A	nr	6.00	57.42	4243.00	645.06	4945.48
OMA16 3200/2500A	nr	6.00	57.42	4597.00	698.16	4654.42
OMA24 2400/3000A	nr	6.00	57.42	6735.00	1018.86	7811.28

4-pole, solenoid 110V DC, draw
out cassetted, frame size

OM8 800/800A	nr	6.00	57.42	2412.00	370.41	2839.83
OM8 1000/1000A	nr	6.00	57.42	2498.00	383.31	2938.73
OM8 1200/1200A	nr	6.00	57.42	2559.00	392.46	3008.88
OMA10 1000/1000A	nr	6.00	57.42	2755.00	421.86	3234.28

	Unit	Labour hours	Net labour (£)	Net material (£)	O'heads /profit (£)	Total (£)
OMA10 1600/1250A	nr	6.00	57.42	2888.00	441.81	3387.23
OMA12 2000/2000A	nr	6.00	57.42	3437.00	524.16	4018.58
OMA12 2500/2000A	nr	6.00	57.42	4044.00	615.21	4716.63
OMA16 3000/2500A	nr	6.00	57.42	4788.00	726.81	5572.23
OMA16 3200/2500A	nr	6.00	57.42	5158.00	782.31	5997.73

Accessories for solenoid breakers, fitted

	Unit	Labour hours	Net labour (£)	Net material (£)	O'heads /profit (£)	Total (£)
Closing control including rectifier, closing contactor and anti-pump relay, panel mounted and wired	nr	3.00	28.71	355.76	57.67	442.14
Trip/close fitted and wired	nr	2.50	23.93	60.05	12.60	96.58
Local/remote switch fitted and wired	nr	2.50	23.93	60.05	12.60	96.58

Accessories for all breakers, fitted

	Unit	Labour hours	Net labour (£)	Net material (£)	O'heads /profit (£)	Total (£)
Under voltage release instantaneous	nr	3.00	28.71	80.44	16.37	125.52
Under voltage release with true lag	nr	3.00	28.71	117.83	21.98	168.52
Shunt trip	nr	3.00	28.71	88.44	17.57	134.72
Auxiliary switch two way (2n/o and 2n/c)	nr	0.75	7.18	62.32	10.42	79.92
Auxiliary switch four way (4n/o and 4n/c)	nr	0.75	7.18	80.44	13.14	100.76
Secondary plugs and sockets	nr	0.33	3.16	84.98	13.22	101.36
Terminal blocks for fixed ACB	nr	0.33	3.16	27.19	4.55	34.90
Carriage auxiliary switch	nr	0.33	3.16	63.45	9.99	76.60
Ammeter (fitted to front cover)	nr	0.50	4.79	53.25	8.71	66.75

HV SWITCHGEAR

	Unit	Labour hours	Net labour (£)	Net material (£)	O'heads /profit (£)	Total (£)
Ammeter and selector switch fitted	nr	0.50	4.79	67.98	10.92	83.69
Castel or Lowe and Fletcher lock	nr	0.50	4.79	49.85	8.20	62.84
Castel or Lowe and Fletcher key	nr	0.00	0.00	16.99	2.55	16.99
Mechanical interlock	nr	1.25	11.96	356.90	55.33	424.19
Hydraulic lifting truck	nr	0.00	0.00	1677.97	251.70	1929.67
Earthing device for use with						
OMA10 TP	nr	2.50	23.93	312.71	50.50	387.14
OMA12 TP	nr	2.50	23.93	328.57	52.88	405.38
OMA16 TP	nr	2.50	23.93	414.68	65.79	504.40
OMA24 TP	nr	2.50	23.93	501.92	78.88	604.73
OMA31 TP	nr	2.50	23.93	581.23	90.77	695.93
Busbar trunking (Ottermill Ltd)						
Rating 125A, length						
3.66m	nr	1.00	9.57	238.29	37.18	285.04
1.83m	nr	0.75	7.18	125.68	19.93	152.79
0.92m	nr	0.50	4.79	62.44	10.08	77.31
Feed unit end	nr	0.50	4.79	58.36	9.47	72.62
Feed unit centre	nr	0.50	4.79	136.09	21.13	162.01
Accessories						
Bend right angle						
flat	nr	0.50	4.79	72.94	11.66	89.39
outside	nr	0.50	4.79	72.94	11.66	89.39
inside	nr	0.50	4.79	72.94	11.66	89.39
Tee						
flat	nr	0.66	6.32	156.08	24.36	186.76
outside	nr	0.66	6.32	156.08	24.36	186.76
inside	nr	0.66	6.32	156.08	24.36	186.76

M & E MEASUREMENT SERVICES

	Unit	Labour hours	Net labour (£)	Net material (£)	O'heads /profit (£)	Total (£)
Crossover						
flat	nr	0.75	7.18	136.09	21.49	164.76
edgewise	nr	0.75	7.18	136.09	21.49	164.76
end cover	nr	0.50	4.79	12.22	2.55	19.56
63A TPN tap-off box	nr	0.50	4.79	55.58	9.06	60.37
line tap	nr	0.50	4.79	3.27	1.21	9.27
Rating 200A, length						
3.66m	nr	1.00	9.57	349.47	53.86	412.90
1.83m	nr	0.75	7.18	183.21	28.56	218.95
0.92m	nr	0.50	4.79	130.89	20.35	156.03
0.3m	nr	0.33	3.16	71.52	11.20	85.88
Feed unit end	nr	0.50	4.79	58.36	9.47	72.62
Feed unit centre	nr	0.50	4.79	136.09	21.13	162.01
Accessories						
Bend right angle						
flat	nr	0.50	4.79	215.00	32.97	252.76
outside	nr	0.50	4.79	215.00	32.97	252.76
inside	nr	0.50	4.79	215.00	32.97	252.76
Tee						
flat	nr	0.66	6.32	215.46	33.27	255.05
outside	nr	0.66	6.32	215.46	33.27	255.05
inside	nr	0.66	6.32	215.46	33.27	255.05
Crossover						
flat	nr	0.75	7.18	136.09	21.49	164.76
edgewise	nr	0.75	7.18	136.09	21.49	164.76
end cover	nr	0.50	4.79	12.22	2.55	19.56
63A TPN tap-off box	nr	0.50	4.79	55.58	9.06	60.37
line tap	nr	0.50	4.79	3.27	1.21	9.27

HV SWITCHGEAR

	Unit	Labour hours	Net labour (£)	Net material (£)	O'heads /profit (£)	Total (£)
Fuse carrier, fuse link and copper link						
For use with integral tap-off unit, rating						
2A	nr	0.17	1.63	1.70	0.50	3.83
6A	nr	0.17	1.63	1.70	0.50	3.83
10A	nr	0.17	1.63	1.70	0.50	3.83
16A	nr	0.17	1.63	1.82	0.52	3.97
20A	nr	0.17	1.63	1.82	0.52	3.97
25A	nr	0.17	1.63	1.98	0.54	4.15
32A	nr	0.17	1.63	1.98	0.54	4.15
For use with 63A external top of boxes, rating						
35A	nr	0.75	7.18	4.22	1.71	13.11
40A	nr	0.75	7.18	4.22	1.71	13.11
50A	nr	0.75	7.18	4.35	1.73	13.26
63A	nr	0.75	7.18	4.96	1.82	13.96
70A	nr	0.75	7.18	6.54	2.06	15.78
80A	nr	0.75	7.18	6.54	2.06	15.78
100A	nr	0.75	7.18	6.54	2.06	15.78
Busbar Chamber System (Bill Switchgear Ltd), rating						
100A, nominal length 550mm	nr	1.80	17.23	130.83	22.21	170.27
100A, nominal length 900mm	nr	2.50	23.93	197.01	33.14	254.08
100A, nominal length 1350mm	nr	3.00	28.71	276.35	45.76	350.82
100A, nominal length 1800mm	nr	3.50	33.50	355.82	58.40	447.72
200A, nominal length 550mm	nr	1.80	17.23	200.93	32.72	250.88
200A, nominal length 900mm	nr	2.50	23.93	289.45	47.00	313.38
200A, nominal length 1350mm	nr	3.00	28.71	395.54	63.64	487.89
200A, nominal length 1800mm	nr	3.50	33.50	501.60	80.27	615.37
400A, nominal length 550mm	nr	1.90	18.18	258.28	41.47	317.93
400A, nominal length 900mm	nr	2.70	25.84	341.19	55.05	422.08
400A, nominal length 1350mm	nr	3.20	30.62	465.51	74.42	570.55
400A, nominal length 1800mm	nr	3.70	35.41	589.89	93.79	719.09
630A, nominal length 900mm	nr	2.70	25.84	504.76	79.59	610.19
630A, nominal length 1350mm	nr	3.20	30.62	745.10	116.36	892.08

M & E MEASUREMENT SERVICES

	Unit	Labour hours	Net labour (£)	Net material (£)	O'heads /profit (£)	Total (£)
630A, nominal length 1800mm	nr	3.70	35.41	857.62	133.95	1026.98
800A, nominal length 900mm	nr	2.70	25.84	840.99	130.02	996.85
800A, nominal length 1350mm	nr	3.20	30.62	1177.91	181.28	1389.81
800A, nominal length 1800mm	nr	3.20	30.62	1454.66	222.79	1708.07
Extension sets for use with the following rated chambers						
200A	nr	1.50	14.36	35.40	7.46	57.22
400A	nr	1.70	16.27	43.01	8.89	68.17
630A	nr	1.70	16.27	65.70	12.30	94.27
800A	nr	1.70	16.27	137.14	23.01	176.42
Pedestal and back plate sets for nominal lengths of busbar chambers						
900mm	nr	0.50	4.79	166.36	25.67	196.82
1350mm	nr	0.50	4.79	196.44	30.18	231.41
1800mm	nr	0.50	4.79	217.33	33.32	255.44
Switchgear mounting sets, switch rating						
63A, busbar rating, all	nr	1.00	9.57	19.00	4.29	28.57
100A, busbar rating, all	nr	1.00	9.57	22.33	4.78	36.68
200A, busbar rating, all	nr	1.00	9.57	22.33	4.78	36.68
315/400A, busbar rating, all	nr	1.00	9.57	24.42	5.10	39.09
630/800A, busbar rating, all	nr	1.00	9.57	28.85	5.76	38.42
Connection sets						
Suitable for the following rated fuse switches						
100A	nr	1.25	11.96	35.77	7.16	54.89
200A	nr	1.25	11.96	40.85	7.92	60.73
315/400A	nr	1.25	11.96	69.81	12.27	94.04
630/800A	nr	1.25	11.96	147.27	23.88	183.11
Busbar clamps						
20 - 60A	nr	0.33	3.16	1.51	0.70	5.37
60 - 100A	nr	0.33	3.16	1.88	0.76	5.80
100 - 200A	nr	0.33	3.16	2.90	0.91	6.97

HV SWITCHGEAR

	Unit	Labour hours	Net labour (£)	Net material (£)	O'heads /profit (£)	Total (£)
Busbar sockets						
100 - 200A	nr	0.33	3.16	8.31	1.72	13.19
200 - 300A	nr	0.33	3.16	11.46	2.19	16.81
300 - 400A	nr	0.33	3.16	14.00	2.57	19.73
Cubicle switch fuses - to BS5419 (Bill Switchgear Ltd), type						
SPN, rating 63A	nr	1.50	14.36	116.05	19.56	149.97
SPN, rating 100A	nr	2.00	19.14	164.08	27.48	210.70
SPN, rating 200A	nr	3.00	28.71	216.09	36.72	281.52
DP, rating 63A	nr	1.50	14.36	124.37	20.81	159.54
DP, rating 100A	nr	2.00	19.14	173.34	28.87	221.35
DP, rating 200A	nr	3.00	28.71	232.08	39.12	299.91
TPN, rating 63A	nr	2.00	19.14	134.58	23.06	176.78
DP, rating 100A	nr	3.00	28.71	184.64	32.00	245.35
DP, rating 200A	nr	3.33	31.87	238.91	40.62	311.40
DP, rating 315A	nr	4.00	38.28	438.16	71.47	547.91
DP, rating 400A	nr	5.00	47.85	541.08	88.34	677.27
DP, rating 630A	nr	5.80	55.51	968.82	153.65	1177.98
DP, rating 800A	nr	6.00	57.42	1151.54	181.34	1390.30
TPSN, rating 100A	nr	3.00	28.71	232.80	39.23	300.74
TPSN, rating 200A	nr	3.33	31.87	280.14	46.80	358.81
TPSN, rating 315A	nr	4.00	38.28	493.06	79.70	611.04
TPSN, rating 400A	nr	5.00	47.85	678.29	108.92	835.06
TPSN, rating 630A	nr	5.80	55.51	1152.46	181.20	1389.17
TPSN, rating 800A	nr	6.00	57.42	1410.55	220.20	1688.17
Switchboard accessories						
Busbar support	nr	0.50	4.79	13.21	2.70	20.70
Riser support	nr	0.50	4.79	7.91	1.91	14.61
Riser support clamp	nr	0.25	2.39	4.09	0.97	7.45
Busbar spacer	nr	0.17	1.63	0.95	0.39	2.97
Riser spacer	nr	0.17	1.63	0.89	0.38	2.90
Packer	nr	0.08	0.77	1.27	0.31	2.35

M & E MEASUREMENT SERVICES

	Unit	Labour hours	Net labour (£)	Net material (£)	O'heads /profit (£)	Total (£)
Terminal supports and gaiters						
support 100A connections	nr	0.33	3.16	2.58	0.86	6.60
gaiter 100A connections	nr	0.33	3.16	2.81	0.90	6.87
support 200A connections	nr	0.33	3.16	2.68	0.88	6.72
gaiter 200A connections	nr	0.33	3.16	3.09	0.94	7.19
support 300/400A connections	nr	0.33	3.16	3.84	1.05	8.05
gaiter 300/400A connections	nr	0.33	3.16	4.66	1.17	8.99
support 600/800A connections	nr	0.33	3.16	8.74	1.78	13.68
gaiter 600/800A connections	nr	0.33	3.16	8.37	1.73	13.26

Surface mounting fuse switches to BS5419 (Bill Switchgear Ltd), type

	Unit	Labour hours	Net labour (£)	Net material (£)	O'heads /profit (£)	Total (£)
SPN, rating 63A	nr	1.50	14.36	135.53	22.48	172.37
SPN, rating 100A	nr	2.00	19.14	196.29	32.31	247.74
SPN, rating 200A	nr	3.00	28.71	278.75	46.12	353.58
DP, rating 63A	nr	1.50	14.36	144.32	23.80	182.48
DP, rating 100A	nr	2.00	19.14	209.01	34.22	262.37
DP, rating 200A	nr	3.00	28.71	296.84	48.83	374.38
TPN, rating 63A	nr	2.00	19.14	164.65	27.57	211.36
TPN, rating 100A	nr	3.00	28.71	220.17	37.33	286.21
TPN, rating 200A	nr	3.33	31.87	324.02	53.38	409.27
TPN, rating 315A	nr	4.00	38.28	579.64	92.69	710.61
TPN, rating 400A	nr	5.00	47.85	684.95	109.92	842.72
TPN, rating 630A	nr	5.80	55.51	1225.34	192.13	1472.98
TPN, rating 800A	nr	6.00	57.42	1437.48	224.23	1719.13
TPSN, rating 100A	nr	3.00	28.71	265.98	44.20	338.89
TPSN, rating 200A	nr	3.33	31.87	394.57	63.97	490.41
TPSN, rating 315A	nr	4.00	38.28	537.45	86.36	662.09
TPSN, rating 400A	nr	5.00	47.85	796.29	126.62	970.76
TPSN, rating 630A	nr	5.80	55.51	1482.68	230.73	1768.92
TPSN, rating 800A	nr	6.00	57.42	1643.75	255.18	1956.35

Cable entry boxes (Bill Switchgear Ltd)

Short cable boxes for copper stranded cables

Type Rating	Unit	Labour hours	Net labour (£)	Net material (£)	O'heads /profit (£)	Total (£)
SPN/DP/TP 200A	nr	3.33	31.87	39.32	10.68	81.87
TP/TPN/TPSN 315A	nr	4.00	38.28	52.45	13.61	104.34

HV SWITCHGEAR

Cable entry boxes (cont'd)	Unit	Labour hours	Net labour (£)	Net material (£)	O'heads /profit (£)	Total (£)
TP/TPN/TPSN 400A	nr	5.00	47.85	52.45	15.04	115.34
TP/TPN/TPSN 630A	nr	5.80	55.51	136.69	28.83	221.03
TP/TPN/TPSN 800A	nr	6.00	57.42	136.69	29.12	223.23

Long cable boxes for solid aluminium cables (Bill Switchgear Ltd)

Type   Rating

SPN/DP/TP/TPN 63A	nr	2.00	19.14	31.00	7.52	57.66
SPN/DP/TP/TPN/TPSN 100A	nr	3.00	28.71	37.49	9.93	76.13
SPN/DP/TP/TPN/TPSN 200A	nr	3.33	31.87	45.99	11.68	89.54
TP/TPN/TPSN 315A	nr	4.00	38.28	62.44	15.11	115.83
TP/TPN/TPSN 400A	nr	5.00	47.85	62.44	16.54	126.83
TP/TPN/TPSN 630A	nr	5.80	55.51	136.69	28.83	221.03
TP/TPN/TPSN 800A	nr	6.00	57.42	136.69	29.12	223.23

Reverse entry cable boxes (Bill Switchgear Ltd)

Type   Rating

SPN/DP/TP/TPN 63A	nr	2.00	19.14	50.32	10.42	79.88
SPN/DP/TP/TPN/TPSN 100A	nr	3.00	28.71	87.24	17.39	133.34
SPN/DP/TP/TPN 200A	nr	3.33	31.87	120.59	22.87	175.33
TP/TPN/TPSN 315A	nr	4.00	38.28	377.37	62.35	478.00
TP/TPN/TPSN 400A	nr	5.00	47.85	425.20	70.96	544.01
TP/TPN/TPSN 630A	nr	5.80	55.51	492.78	82.24	630.53
TP/TPN/TPSN 800A	nr	6.00	57.42	544.25	90.25	691.92

	Unit	Labour hours	Net labour (£)	Net material (£)	O'heads /profit (£)	Total (£)

Y71 LV SWITCHGEAR AND DISTRIBUTION BOARDS

General notes

1. All labour times allow for fixing equipment to backgrounds including drilling, plugging and screwing

2. All labour times on switches and protection equipment allow for standard cable connections

3. No discounts have been applied to any of the prices in this section due to the variance of discounts on different items of equipment

(Dorman Smith Switchgear Ltd)

800A fuse switch, panel board, not including fuse switches

	Unit	Labour hours	Net labour (£)	Net material (£)	O'heads /profit (£)	Total (£)
6 way, TP&N	nr	2.00	19.14	542.49	84.24	645.87
10 way, TP&N	nr	3.00	28.71	856.83	132.83	1018.37

Fuse switches mounted within panel boards

	Unit	Labour hours	Net labour (£)	Net material (£)	O'heads /profit (£)	Total (£)
32A, TP&N	nr	1.00	9.57	56.87	9.97	76.41
63A, SP&N	nr	1.00	9.57	80.14	13.46	89.71
63A, TP&N	nr	1.00	9.57	84.78	14.15	108.50
100A, SP&N	nr	1.00	9.57	102.51	16.81	128.89
100A, TP&N	nr	1.00	9.57	144.67	23.14	177.38

200A, D type, TP&N

	Unit	Labour hours	Net labour (£)	Net material (£)	O'heads /profit (£)	Total (£)
200A, TP&N	nr	1.10	10.53	177.52	28.21	216.26
315A, TP&N	nr	1.10	10.53	323.78	50.15	384.46
400A, TP&N	nr	1.40	13.40	391.28	60.70	465.38
630A, TP&N	nr	1.60	15.31	702.14	107.62	825.07
800A, TP&N	nr	1.80	17.23	793.57	121.62	932.42

LV SWITCHGEAR

	Unit	Labour hours	Net labour (£)	Net material (£)	O'heads /profit (£)	Total (£)
400A MCCB panel board, not including MCCBs						
6 way, TP&N	nr	1.50	14.36	233.00	37.10	284.46
12 way, TP&N	nr	2.20	21.05	283.85	45.73	350.63
800A MCCB panel board, not including MCCBs						
6 way, TP&N	nr	1.65	15.79	278.59	44.16	338.54
12 way, TP&N	nr	2.35	22.49	368.37	58.63	449.49
MCCB panel board accessories						
SP MCCB blanking piece	nr	0.10	0.96	0.73	0.25	1.94
SP MCCB terminal shroud	nr	0.10	0.96	2.26	0.48	3.70
busbar interconnection kit for stacking two 400A MCCB panel boards	nr	1.50	14.36	22.57	5.54	36.93
busbar interconnection kit for stacking two 800A MCCB panel boards		1.60	15.31	31.60	7.04	46.91
door lock with keys	nr	0.50	4.79	4.51	1.40	9.30
Moulded case circuit breakers (MCCBs), mounted within panel boards						
Single pole MCCBs						
25A and 32A	nr	0.50	4.79	41.44	6.93	46.23
40A and 50A	nr	0.50	4.79	42.00	7.02	46.79
63A and 80A	nr	0.50	4.79	42.57	7.10	54.46
100A	nr	0.50	4.79	43.14	7.19	55.12
125A and 160A	nr	0.50	4.79	87.98	13.92	92.77
200A	nr	0.50	4.79	100.47	15.79	121.05
225A	nr	0.50	4.79	116.36	18.17	121.15
250A	nr	0.50	4.79	122.61	19.11	146.51
Triple pole MCCBs						
25A and 32A	nr	0.70	6.70	77.77	12.67	84.47
40A and 50A	nr	0.70	6.70	77.77	12.67	84.47
63A	nr	0.70	6.70	80.60	13.09	100.39

	Unit	Labour hours	Net labour (£)	Net material (£)	O'heads /profit (£)	Total (£)
80A	nr	0.70	6.70	85.14	13.78	105.62
100A	nr	0.70	6.70	88.55	14.29	109.54
125A	nr	0.70	6.70	122.04	19.31	148.05
160A	nr	0.70	6.70	143.04	22.46	172.20
200A	nr	0.70	6.70	164.61	25.70	197.01
250A	nr	0.70	6.70	190.72	29.61	227.03

Triple pole and neutral

	Unit	Labour hours	Net labour	Net material	O'heads /profit	Total
25A and 32A	nr	0.80	7.66	129.42	20.56	137.08
40A and 50A	nr	0.80	7.66	129.42	20.56	137.08
63A	nr	0.80	7.66	133.39	21.16	162.21
80A and 100A	nr	0.80	7.66	136.23	21.58	143.89
125A	nr	0.80	7.66	192.98	30.10	230.74
160A	nr	0.80	7.66	204.35	31.80	243.81
200A	nr	0.80	7.66	249.76	38.61	296.03
250A	nr	0.80	7.66	298.57	45.93	352.16

MCB distribution boards fixed to backgrounds, excluding MCBs

SP&N with switch, disconnector/ direct connection

	Unit	Labour hours	Net labour	Net material	O'heads /profit	Total
6 way	nr	1.40	13.40	20.23	5.04	38.67
9 way	nr	1.90	18.18	25.55	6.56	50.29
12 way	nr	2.40	22.97	31.94	8.24	63.15
15 way	nr	2.80	26.80	38.33	9.77	74.90

SP&N - RCCB incomer selection distribution board, fixed to backgrounds, excluding MCBs

	Unit	Labour hours	Net labour	Net material	O'heads /profit	Total
4 way	nr	1.40	13.40	20.23	5.04	38.67
7 way	nr	1.50	14.36	25.55	5.99	45.90
10 way	nr	2.40	22.97	31.94	8.24	63.15
13 way	nr	2.70	25.84	38.33	9.63	73.80

Accessories for SP&N distribution boards

	Unit	Labour hours	Net labour	Net material	O'heads /profit	Total
direct connection lit	nr	1.00	9.57	5.32	2.23	14.89
cylinder door lock	nr	0.10	0.96	4.51	0.82	6.29
door padlock	nr	0.02	0.19	11.92	1.82	13.93

LV SWITCHGEAR

	Unit	Labour hours	Net labour (£)	Net material (£)	O'heads /profit (£)	Total (£)
front plate blanking piece (12 pieces)	nr	0.05	0.48	1.51	0.30	1.99
12 way earth bar	nr	0.20	1.91	2.13	0.61	4.65
terminal blanking kit	nr	0.10	0.96	1.22	0.33	2.51
TP&N distribution boards fixed to backgrounds, excluding MCBs						
4 way	nr	2.00	19.14	67.40	12.98	99.52
6 way	nr	2.20	21.05	76.97	14.70	112.72
8 way	nr	2.40	22.97	86.24	16.38	125.59
12 way	nr	2.60	24.88	105.64	19.58	150.10
16 way	nr	2.80	26.80	152.54	26.90	206.24
Accessories for TP&N distribution boards						
cylinder door lock	nr	0.10	0.96	4.51	0.82	6.29
door padlock	nr	0.02	0.19	11.92	1.82	12.11
front plate blanking piece (12 pieces)	nr	0.05	0.48	1.51	0.30	2.29
TP&N to SP&N conversion kit	nr	1.20	11.48	14.87	3.95	30.30
earth isolation kit	nr	0.50	4.79	6.23	1.65	12.67
stacking kit	nr	0.50	4.79	21.33	3.92	30.04
integral modular automation mounting	nr	0.40	3.83	23.70	4.13	31.66
spare end plate	nr	0.05	0.48	5.93	0.96	7.37
Incomer options for integral TP&N boards						
Direct connection						
70mm2 cable	nr	0.30	2.87	6.44	1.40	9.31
150mm2 cable	nr	0.40	3.83	10.03	2.08	15.94
RCCB connection kit	nr	0.40	3.83	23.70	4.13	27.53
Fuse-switch connection kit	nr	0.40	3.83	46.81	7.60	50.64

M & E MEASUREMENT SERVICES

	Unit	Labour hours	Net labour (£)	Net material (£)	O'heads /profit (£)	Total (£)
Incoming switch and protection devices for SP&N and TP&N distribution boards						
100A switch disconnects						
2 pole	nr	0.60	5.74	10.60	2.45	18.79
3 pole	nr	0.80	7.66	26.34	5.10	39.10
RCCBs						

Rating	Frame	Pole	Unit	Labour hours	Net labour (£)	Net material (£)	O'heads /profit (£)	Total (£)
40A	30mA	4	nr	0.40	3.83	45.25	7.36	49.08
63A	30mA	2	nr	0.40	3.83	47.19	7.65	51.02
63A	30mA	4	nr	0.40	3.83	62.21	9.91	75.95
100A	30mA	2	nr	0.35	3.35	62.21	9.83	75.39
100A	30mA	4	nr	0.40	3.83	141.39	21.78	167.00
40A	100mA	4	nr	0.40	3.83	39.58	6.51	49.92
63A	100mA	2	nr	0.35	3.35	43.26	6.99	53.60
63A	100mA	4	nr	0.40	3.83	45.25	7.36	56.44
100A	100mA	2	nr	0.35	3.35	55.62	8.85	67.82
100A	100mA	4	nr	0.40	3.83	93.83	14.65	112.31
63A	300mA	2	nr	0.35	3.35	39.33	6.40	49.08
63A	300mA	4	nr	0.40	3.83	42.58	6.96	53.37
100A	300mA	4	nr	0.40	3.83	86.13	13.49	103.45

Enclosed fuse switch		Unit	Labour hours	Net labour (£)	Net material (£)	O'heads /profit (£)	Total (£)
100A	TP	nr	0.40	3.83	161.45	24.79	190.07
100A	TP&N	nr	0.45	4.31	170.27	26.19	174.58
100A	4 pole	nr	0.45	4.31	183.04	28.10	215.45
200A	TP	nr	0.40	3.83	253.76	38.64	296.23
200A	TP&N	nr	0.45	4.31	259.21	39.53	303.05
200A	4 pole	nr	0.45	4.31	285.54	43.48	333.33

LV SWITCHGEAR

	Unit	Labour hours	Net labour (£)	Net material (£)	O'heads /profit (£)	Total (£)
MCBs fixed to DIN rails mounted within a distribution board						
Single pole MCBs						
Type 2 M9						
6A, 50A, 63A	nr	0.17	1.63	6.38	1.20	8.01
10A to 40A	nr	0.17	1.63	5.97	1.14	7.60
Type 3 M9						
6A, 50A, 63A	nr	0.17	1.63	6.51	1.22	8.14
10A to 40A	nr	0.17	1.63	6.09	1.16	7.72
D characteristic						
Type 4 6kA						
6A, 40A, 50A, 63A	nr	0.17	1.63	7.00	1.29	8.63
10A to 32A	nr	0.17	1.63	6.54	1.23	8.17
Double pole MCBs						
Type 2 M9						
6A, 50A, 63A	nr	0.20	1.91	18.54	3.07	20.45
10A to 40A	nr	0.20	1.91	17.37	2.89	19.28
Type 3 M9						
6A, 50A, 63A	nr	0.20	1.91	18.90	3.12	20.81
10A to 40A	nr	0.20	1.91	17.67	2.94	19.58
D characteristic						
Type 4 6kA						
6A, 40A, 50A, 63A	nr	0.20	1.91	20.32	3.33	22.23
10A to 32A	nr	0.20	1.91	19.01	3.14	20.92

M & E MEASUREMENT SERVICES

	Unit	Labour hours	Net labour (£)	Net material (£)	O'heads /profit (£)	Total (£)
Triple pole MCBs						
6A, 50A, 63A	nr	0.23	2.20	27.64	4.48	29.84
10A to 40A	nr	0.23	2.20	25.98	4.23	28.18
Type 3 M9						
6A, 50A, 63A	nr	0.23	2.20	27.87	4.51	30.07
10A to 40A	nr	0.23	2.20	26.21	4.26	28.41
D characteristic						
Type 4 6kA						
6A, 40A, 50A, 63A	nr	0.23	2.20	29.97	4.83	32.17
10A to 32A	nr	0.23	2.20	28.17	4.56	30.37
RCBOs fixed to DIN rails mounted within a distribution board						
16A/30mA, 2 pole	nr	0.17	1.63	36.46	5.71	43.80
20A/30mA, 2 pole	nr	0.17	1.63	36.46	5.71	43.80
32A/30mA, 2 pole	nr	0.17	1.63	36.46	5.71	43.80

CONTACTORS AND STARTERS

	Unit	Labour hours	Net labour (£)	Net material (£)	O'heads /profit (£)	Total (£)

Y72 CONTACTORS AND STARTERS

General notes

1. All labour times allow for fixing equipment to backgrounds including drilling, plugging and screwing

2. Labour times on contactors and the like allow for standard cable connections

3. No discounts have been applied to any of the prices in this section

(Dorman Smith Switchgear Ltd)

Contactors 240V, mounted on DIN rails within MCB distribution boards

Without manual override
Rating   Poles
A   (normally open)

Rating	Poles	Unit	Labour hours	Net labour	Net material	O'heads/profit	Total
20	2	nr	0.20	1.91	13.32	2.28	17.51
20	3	nr	0.25	2.39	15.95	2.75	21.09
20	4	nr	0.33	3.16	17.49	3.10	23.75
40	3	nr	0.25	2.39	25.40	4.17	31.96
40	4	nr	0.33	3.16	30.00	4.97	38.13
63	3	nr	0.25	2.39	46.32	7.31	56.02
63	4	nr	0.33	3.16	50.86	8.10	62.12

With manual override

Rating	Poles	Unit	Labour hours	Net labour	Net material	O'heads/profit	Total
20	2	nr	0.20	1.91	21.87	3.57	23.78
20	3	nr	0.25	2.39	26.93	4.40	29.32
20	4	nr	0.33	3.16	39.44	6.39	42.60
40	3	nr	0.25	2.39	42.16	6.68	44.55
40	4	nr	0.33	3.16	49.33	7.87	52.49

M & E MEASUREMENT SERVICES

	Unit	Labour hours	Net labour (£)	Net material (£)	O'heads /profit (£)	Total (£)
Switches rating 16A 250V AC contact pairs						
1 normally open	nr	0.33	3.16	3.75	1.04	7.95
1 changeover centre off	nr	0.33	3.16	4.00	1.07	8.23
2 normally open	nr	0.33	3.16	4.33	1.12	8.61
1 normally open and neon	nr	0.33	3.16	4.33	1.12	8.61
2 normally open and neon	nr	0.33	3.16	5.04	1.23	9.43
1-2 way changeover	nr	0.33	3.16	4.33	1.12	8.61
2-2 way changeover	nr	0.33	3.16	8.93	1.81	13.90
Push buttons, rating 6A 250V AC, contact pairs						
1 normally open	nr	0.33	3.16	3.88	1.06	8.10
1 normally open and 1 normally closed	nr	0.33	3.16	4.33	1.12	8.61
1 normally open and neon	nr	0.33	3.16	5.90	1.36	10.42
Relays						
thermostatic relay	nr	0.75	7.18	103.98	16.67	111.16
photo-sensitive relay	nr	0.75	7.18	71.36	11.78	78.54
Relay system						
RCD relay	nr	0.50	4.79	107.74	16.88	129.41
100A Toroidal CT	nr	0.50	4.79	20.32	3.77	28.88
250A Toroidal CT	nr	0.50	4.79	48.29	7.96	61.04
500A Toroidal CT	nr	0.55	5.26	69.88	11.27	86.41
1600A Stadium Toroidal CT	nr	0.65	6.22	119.42	18.85	144.49
1600A Toroidal CT	nr	0.65	6.22	119.42	18.85	144.49

CONTACTORS AND STARTERS

	Unit	Labour hours	Net labour (£)	Net material (£)	O'heads /profit (£)	Total (£)

Panel or chassis mounting step down transformers

Single phase 40-100Hz
Air cooled, continuously rated
Simmonds Brothers
Type FF - frames with 25cm
minimum flying leads to input
and output

Capacity

	Unit	Labour hours	Net labour (£)	Net material (£)	O'heads /profit (£)	Total (£)
30VA	nr	0.50	4.79	19.55	3.65	27.99
50VA	nr	0.50	4.79	20.26	3.76	28.81
75VA	nr	0.50	4.79	22.27	4.06	31.12
100VA	nr	0.50	4.79	23.69	4.27	32.75
150VA	nr	0.50	4.79	26.66	4.72	36.17
250VA	nr	0.50	4.79	32.22	5.55	42.56
500VA	nr	0.50	4.79	44.77	7.43	56.99
750VA	nr	0.50	4.79	54.72	8.93	68.44
1000VA	nr	0.50	4.79	61.72	9.98	66.51
1500VA	nr	0.50	4.79	83.51	13.25	88.30
2000VA	nr	0.50	4.79	101.75	15.98	106.54

Type SF - shrouds with 25cm
minimum flying leads to input
and output

Capacity

	Unit	Labour hours	Net labour (£)	Net material (£)	O'heads /profit (£)	Total (£)
30VA	nr	0.50	4.79	19.55	3.65	27.99
50VA	nr	0.50	4.79	22.03	4.02	30.84
75VA	nr	0.50	4.79	23.34	4.22	32.35
100VA	nr	0.50	4.79	25.24	4.50	34.53
150VA	nr	0.50	4.79	27.60	4.86	37.25
250VA	nr	0.50	4.79	34.35	5.87	45.01

M & E MEASUREMENT SERVICES

	Unit	Labour hours	Net labour (£)	Net material (£)	O'heads /profit (£)	Total (£)
Type FT - frames with top mounted terminals to input and output						
Capacity						
30VA	nr	0.50	4.79	20.02	3.72	28.53
50VA	nr	0.50	4.79	20.91	3.85	29.55
75VA	nr	0.50	4.79	22.74	4.13	31.66
100VA	nr	0.50	4.79	24.05	4.33	33.17
150VA	nr	0.50	4.79	27.13	4.79	36.71
250VA	nr	0.50	4.79	32.82	5.64	43.25
500VA	nr	0.50	4.79	45.86	7.60	58.25
750VA	nr	0.50	4.79	55.80	9.09	69.68
1000VA	nr	0.50	4.79	62.90	10.15	67.69
1500VA	nr	0.50	4.79	85.17	13.49	89.96
2000VA	nr	0.50	4.79	103.78	16.29	108.57
Type ST - shrouds with side mounted terminals to input and output						
Capacity						
30VA	nr	0.50	4.79	20.37	3.77	28.93
50VA	nr	0.50	4.79	22.74	4.13	31.66
75VA	nr	0.50	4.79	24.16	4.34	33.29
100VA	nr	0.50	4.79	26.06	4.63	35.48
150VA	nr	0.50	4.79	28.66	5.02	38.47
250VA	nr	0.50	4.79	35.66	6.07	46.52
All insulated transformers for portable tools (Simmonds Bros Ltd)						
Single socket/plug **Rating**						
250/400 VA	nr	0.00	0.00	70.96	10.64	81.60
500/850 VA	nr	0.00	0.00	85.41	12.81	98.22
750/1250 VA	nr	0.00	0.00	111.58	16.74	128.32
1250/2000 VA	nr	0.00	0.00	128.40	19.26	128.40

CONTACTORS AND STARTERS

	Unit	Labour hours	Net labour (£)	Net material (£)	O'heads /profit (£)	Total (£)
Twin sockets/plugs Rating						
250/400VA	nr	0.00	0.00	83.98	12.60	83.98
Twin sockets/plugs Rating						
500/850VA	nr	0.00	0.00	98.44	14.77	98.44
750/1250VA	nr	0.00	0.00	124.61	18.69	124.61
1250/2000VA	nr	0.00	0.00	141.43	21.21	141.43

Metal clad transformers for portable tools (Simmonds Bros Ltd)

Type A (wallmounting with conduit holes) and Type F (free standing with conduit holes)

Portable tool use Capacity (VA)	Continuous to BS3535 Capacity (VA)	Unit	Labour hours	Net labour (£)	Net material (£)	O'heads /profit (£)	Total (£)
250	150	nr	0.00	0.00	45.84	6.88	52.72
400	250	nr	0.00	0.00	52.95	7.94	60.89
850	500	nr	0.00	0.00	67.28	10.09	77.37
1250	750	nr	0.00	0.00	77.23	11.58	77.23
1750	1000	nr	0.00	0.00	91.57	13.74	91.57
2500	1500	nr	0.00	0.00	113.12	16.97	113.12
3500	2000	nr	0.00	0.00	134.45	20.17	134.45
5000	3000	nr	0.00	0.00	183.12	27.47	183.12
7000	4000	nr	0.00	0.00	220.20	33.03	220.20
8500	5000	nr	0.00	0.00	266.16	39.92	266.16
10,000	6000	nr	0.00	0.00	350.02	52.50	350.02
14,000	8000	nr	0.00	0.00	421.57	63.24	421.57
17,500	10,000	nr	0.00	0.00	556.48	83.47	639.95

M & E MEASUREMENT SERVICES

		Unit	Labour hours	Net labour (£)	Net material (£)	O'heads /profit (£)	Total (£)
Type D portable with TRS input cable and output sockets and plugs							
Portable tool use Capacity (VA)	Continuous to BS3535 Capacity (VA)						
250	150	nr	0.00	0.00	62.90	9.43	72.33
400	250	nr	0.00	0.00	70.12	10.52	80.64
850	150	nr	0.00	0.00	84.46	12.67	97.13
1250	750	nr	0.00	0.00	98.67	14.80	98.67
1750	1000	nr	0.00	0.00	113.00	16.95	113.00
2500	1500	nr	0.00	0.00	135.00	20.25	135.00
3500	2000	nr	0.00	0.00	170.22	25.53	170.22
5000	3000	nr	0.00	0.00	223.16	33.47	223.16
7000	4000	nr	0.00	0.00	288.90	43.33	288.90
8500	5000	nr	0.00	0.00	348.95	52.34	348.95
10,000	6000	nr	0.00	0.00	433.05	64.96	433.05

LUMINAIRES AND LAMPS

	Unit	Labour hours	Net labour (£)	Net material (£)	O'heads /profit (£)	Total (£)

Y73 LUMINAIRES AND LAMPS
(Philips lighting)

General notes

1. All labour times include fixing luminaires and accessories to backgrounds, including all drilling, supports and bracketry

2. All labour times on luminaires allow for standard cable connections and terminators

3. No discounts have been incorporated to any of the prices in this section due to the variance of discounts for each luminaire

4. All labour times include installing luminaires up to a maximum height of 4.5m but not erecting trestles or scaffolding. Add 10% to times for heights 4.5-7m

Commercial/industrial fluorescent luminaires

Streampak multi-purpose batten fittings, switch start complete standard white fluorescent lamps

No of tubes	Length mm		Unit	Labour hours	Net labour (£)	Net material (£)	O'heads /profit (£)	Total (£)
1 x 18W	600		nr	0.50	4.79	15.16	2.99	22.94
2 x 18W	600		nr	0.58	5.55	26.07	4.74	36.36
1 x 36W	1200		nr	0.75	7.18	18.56	3.86	29.60
2 x 36W	1200		nr	0.83	7.94	37.30	6.79	52.03
1 x 58W	1500		nr	1.00	9.57	21.62	4.68	35.87
2 x 58W	1500		nr	1.25	11.96	42.30	8.14	62.40
1 x 70W	1800		nr	1.10	10.53	26.12	5.50	42.15

		Unit	Labour hours	Net labour (£)	Net material (£)	O'heads /profit (£)	Total (£)
2 x 70W	1800	nr	1.33	12.73	46.33	8.86	67.92
1 x 100W	2400	nr	1.20	11.48	37.45	7.34	56.27
2 x 100W	2400	nr	1.50	14.36	63.65	11.70	89.71

Streampak multi-purpose batten emergency system incorporating a self-contained 3 hour emergency pack

No of tubes	Length mm						
1 x 18W	600	nr	0.62	5.93	115.16	18.16	139.25
2 x 18W	600	nr	0.68	6.51	126.07	19.89	152.47
1 x 36W	1200	nr	0.75	7.18	118.56	18.86	144.60
2 x 36W	1200	nr	0.83	7.94	137.30	21.79	167.03
1 x 58W	1500	nr	1.00	9.57	121.62	19.68	150.87
2 x 58W	1500	nr	1.25	11.96	142.30	23.14	177.40
1 x 70W	1800	nr	1.10	10.53	126.12	20.50	157.15
2 x 70W	1800	nr	1.33	12.73	146.38	23.87	182.98
1 x 100W	2400	nr	1.20	11.48	137.45	22.34	171.27
2 x 100W	2400	nr	1.50	14.36	163.65	26.70	204.71

All the following attachments are for use with Philips Streampak multi-purpose batten luminaires

One lamp opal diffuser, length

600mm	nr	0.13	1.24	7.64	1.33	10.21
1200mm	nr	0.13	1.24	10.24	1.72	13.20
1500mm	nr	0.20	1.91	11.53	2.02	15.46
1800mm	nr	0.20	1.91	13.26	2.28	17.45
2400mm	nr	0.20	1.91	19.59	3.22	24.72

Two lamp opal diffuser, length

600mm	nr	0.13	1.24	8.52	1.46	11.22
1200mm	nr	0.13	1.24	19.99	3.18	24.41
1500mm	nr	0.20	1.91	24.86	4.02	30.79
1800mm	nr	0.20	1.91	28.89	4.62	35.42
2400mm	nr	0.20	1.91	38.16	6.01	46.08

LUMINAIRES AND LAMPS

	Unit	Labour hours	Net labour (£)	Net material (£)	O'heads /profit (£)	Total (£)
One or two lamp prismatic controllers, length						
600mm	nr	0.13	1.24	8.95	1.53	11.72
1200mm	nr	0.13	1.24	13.26	2.17	16.67
1500mm	nr	0.20	1.91	14.64	2.48	19.03
1800mm	nr	0.20	1.91	18.24	3.02	23.17
2400mm	nr	0.20	1.91	23.62	3.83	29.36
Two lamp prismatic controllers, length						
600mm	nr	0.13	1.24	12.51	2.06	13.75
1200mm	nr	0.13	1.24	22.68	3.59	27.51
1500mm	nr	0.20	1.91	28.45	4.55	34.91
1800mm	nr	0.20	1.91	35.85	5.66	43.42
2400mm	nr	0.20	1.91	41.29	6.48	43.20
Open end metal trough reflectors, slotted 1 or 2 lamp, length						
1200mm	nr	0.13	1.24	12.59	2.07	15.90
1500mm	nr	0.20	1.91	13.15	2.26	17.32
1800mm	nr	0.20	1.91	14.87	2.52	19.30
2400mm	nr	0.20	1.91	22.56	3.67	28.14
Wire guard for use with trough reflectors, length						
1500mm	nr	0.20	1.91	27.75	4.45	34.11
1800mm	nr	0.20	1.91	33.52	5.31	40.74
2400mm	nr	0.20	1.91	38.63	6.08	46.62
Open end metal angle reflectors 1 or 2 lamp, length						
1500mm	nr	0.20	1.91	17.07	2.85	21.83
1800mm	nr	0.20	1.91	18.41	3.05	23.37
2400mm	nr	0.20	1.91	20.71	3.39	26.01

M & E MEASUREMENT SERVICES

	Unit	Labour hours	Net labour (£)	Net material (£)	O'heads /profit (£)	Total (£)
Wire guard for use with angle reflectors, length						
1500mm	nr	0.20	1.91	26.44	4.25	32.60
1800mm	nr	0.20	1.91	31.88	5.07	38.86
2400mm	nr	0.20	1.91	36.84	5.81	44.56
Streampak accessories						
single lamp holder	nr	0.25	2.39	2.54	0.74	5.67
twin lamp holder	nr	0.27	2.58	2.88	0.82	6.28
single diffuser/controller bracket	nr	0.17	1.63	0.58	0.33	2.54
twin diffuser/controller bracket	nr	0.23	2.20	1.33	0.53	3.53
single batten end cap	nr	0.13	1.24	0.74	0.30	1.98
twin batten end cap	nr	0.13	1.24	0.74	0.30	1.98
single diffuser end cap	nr	0.13	1.24	0.74	0.30	2.28
twin diffuser end cap	nr	0.13	1.24	1.02	0.34	2.60
Philips TCS314 range of surface mounted luminaires complete with standard control gear						
3 x 18W 600mm, switchstart	nr	1.00	9.57	108.12	17.65	117.69
4 x 18W 600mm, switchstart	nr	1.00	9.57	117.90	19.12	146.59
1 x 36W 1200mm, switchstart	nr	1.13	10.81	73.49	12.64	96.94
2 x 36W 1200mm, switchstart	nr	1.40	13.40	81.90	14.29	109.59
1 x 58W 1500mm, switchstart	nr	1.40	13.40	85.12	14.78	113.30
2 x 58W 1500mm, switchstart	nr	1.40	13.40	89.52	15.44	118.36
Emergency fittings (3 hour self-contained, maintained system)						
3 x 18W, 600mm	nr	1.00	9.57	218.12	34.15	261.84
4 x 18W, 600mm	nr	1.00	9.57	217.00	33.99	260.56
1 x 36W, 1200mm	nr	1.13	10.81	173.49	27.64	211.94
2 x 36W, 1200mm	nr	1.13	10.81	181.90	28.91	221.62
1 x 58W, 1500mm	nr	1.40	13.40	185.12	29.78	228.30
2 x 58W, 1500mm	nr	1.40	13.40	189.52	30.44	202.92

LUMINAIRES AND LAMPS

	Unit	Labour hours	Net labour (£)	Net material (£)	O'heads /profit (£)	Total (£)
Diffusers and reflectors for the TCS314 range						
Prismatic diffusers for						
3 or 4 x 18W, 600mm body	nr	0.13	1.24	36.11	5.60	37.35
1 x 36W, 1200mm body	nr	0.13	1.24	24.96	3.93	30.13
2 x 36W, 1200mm body	nr	0.15	1.44	34.21	5.35	41.00
1 x 58W, 1500mm body	nr	0.15	1.44	28.69	4.52	34.65
2 x 58W, 1500mm body	nr	0.17	1.63	39.51	6.17	47.31
CIBSE LG3, category 3, mirror reflector for						
3 x 18W, 600mm body	nr	0.13	1.24	68.72	10.49	80.45
4 x 18W, 600mm body	nr	0.13	1.24	72.60	11.08	73.84
1 x 36W, 1200mm body	nr	0.13	1.24	54.37	8.34	63.95
2 x 36W, 1200mm body	nr	0.15	1.44	69.32	10.61	81.37
1 x 58W, 1500mm body	nr	0.15	1.44	59.86	9.20	70.50
2 x 58W, 1500mm body	nr	0.17	1.63	74.90	11.48	88.01
CIBSE LG3, category 2, mirror reflector for						
3 x 18W, 600mm body	nr	0.13	1.24	77.65	11.83	90.72
4 x 18W, 600mm body	nr	0.13	1.24	81.99	12.48	95.71
1 x 36W, 1200mm body	nr	0.13	1.24	63.07	9.65	73.96
2 x 36W, 1200mm body	nr	0.15	1.44	78.33	11.97	91.74
1 x 58W, 1500mm body	nr	0.15	1.44	68.95	10.56	80.95
2 x 58W, 1500mm body	nr	0.17	1.63	83.96	12.84	98.43
Philips TBS300 range of measured modular luminaires complete with switch start control gear and brackets						
2 x 18W, 600 x 300mm	nr	0.50	4.79	78.90	12.55	96.24
3 x 18W, 600 x 600mm	nr	0.50	4.79	94.45	14.89	114.13
4 x 18W, 600 x 600mm	nr	0.50	4.79	95.66	15.07	115.52
1 x 36W, 1200 x 300mm	nr	0.52	4.98	55.40	9.06	69.44
2 x 36W, 1200 x 300mm	nr	0.52	4.98	76.92	12.29	94.19
3 x 36W, 1200 x 600mm	nr	0.56	5.36	109.38	17.21	131.95

	Unit	Labour hours	Net labour (£)	Net material (£)	O'heads /profit (£)	Total (£)
4 x 36W, 1200 x 600mm	nr	0. 56	5. 36	130. 41	20. 37	135. 77
1 x 58W, 1500 x 300mm	nr	0. 60	5. 74	59. 80	9. 83	65. 54
2 x 58W, 1500 x 300mm	nr	0. 62	5. 93	83. 93	13. 48	89. 86

Emergency fittings incorporating
a 3 hour self-contained maintained
system

2 x 18W, 600 x 300mm	nr	0. 63	6. 03	178. 90	27. 74	212. 67
3 x 18W, 600 x 600mm	nr	0. 63	6. 03	194. 45	30. 07	230. 55
4 x 18W, 600 x 600mm	nr	0. 63	6. 03	195. 66	30. 25	231. 94
1 x 36W, 1200 x 300mm	nr	0. 67	6. 41	155. 40	24. 27	186. 08
2 x 36W, 1200 x 300mm	nr	0. 67	6. 41	176. 92	27. 50	210. 83
3 x 36W, 1200 x 600mm	nr	0. 72	6. 89	209. 38	32. 44	248. 71
4 x 36W, 1200 x 600mm	nr	0. 72	6. 89	230. 41	35. 59	272. 89
1 x 58W, 1500 x 300mm	nr	0. 76	7. 27	159. 80	25. 06	192. 13
2 x 58W, 1500 x 600mm	nr	0. 78	7. 46	283. 93	43. 71	291. 39

Diffusers and reflectors for the
TBS300 range

Prismatic diffusers for

3 or 4 x 18W body	nr	0. 13	1. 24	36. 11	5. 60	42. 95
1 or 2 x 36W body	nr	0. 13	1. 24	34. 21	5. 32	40. 77
3 or 4 x 36W body	nr	0. 17	1. 63	54. 97	8. 49	65. 09
1 or 2 x 58W body	nr	0. 17	1. 63	39. 51	6. 17	47. 31

Standard mirror reflector for

3 x 18W body	nr	0. 13	1. 24	35. 80	5. 56	42. 60
4 x 18W body	nr	0. 13	1. 24	38. 85	6. 01	46. 10
2 x 36W body	nr	0. 15	1. 44	37. 48	5. 84	44. 76
3 x 36W body	nr	0. 17	1. 63	69. 10	10. 61	81. 34
4 x 36W body	nr	0. 19	1. 82	73. 56	11. 31	86. 69
2 x 58W body	nr	0. 19	1. 82	43. 39	6. 78	51. 99

CIBSE LG3, category 3, mirror
reflectors for

2 x 18W body	nr	0. 13	1. 24	51. 37	7. 89	60. 50
3 x 18W body	nr	0. 13	1. 24	68. 72	10. 49	80. 45
4 x 18W body	nr	0. 13	1. 24	72. 60	11. 08	84. 92
1 x 36W body	nr	0. 13	1. 24	54. 32	8. 33	63. 89

Mirror reflectors (cont'd)	Unit	Labour hours	Net labour (£)	Net material (£)	O'heads /profit (£)	Total (£)
2 x 36W body	nr	0.15	1.44	69.32	10.61	81.37
3 x 36W body	nr	0.17	1.63	96.99	16.99	98.62
4 x 36W body	nr	0.19	1.82	106.40	16.23	124.45
1 x 58W body	nr	0.15	1.44	59.86	9.20	61.30
2 x 58W body	nr	0.19	1.82	74.90	11.51	88.23
CIBSE LG3 category 2, mirror reflectors for						
2 x 18W body	nr	0.13	1.24	62.40	9.55	73.19
3 x 18W body	nr	0.13	1.24	77.65	11.83	90.72
4 x 18W body	nr	0.13	1.24	81.97	12.48	95.69
1 x 36W body	nr	0.13	1.24	63.07	9.65	73.96
2 x 36W body	nr	0.15	1.44	78.33	11.97	91.74
3 x 36W body	nr	0.17	1.63	112.41	17.11	131.15
4 x 36W body	nr	0.19	1.82	127.80	19.44	149.06
1 x 58W body	nr	0.15	1.44	68.95	10.56	80.95
2 x 58W body	nr	0.19	1.82	83.96	12.87	98.65
TBS300 accessories						
lampholder, fitted	nr	0.25	2.39	3.31	0.85	6.55
'U' spring retaining clip for prismatic diffuser	nr	0.17	1.63	1.75	0.51	3.89
'U' spring retaining clip for mirror reflectors	nr	0.17	1.63	1.44	0.46	3.53
brackets (set of 4) for concealed tee ceilings	nr	0.33	3.16	8.98	1.82	13.96
Philips Pacific, vandal proof luminaire, surface mounted complete with a polycarbonate diffuser and lockable vandal resistant clips						
1 x 18W 600mm, switchstart	nr	0.50	4.79	48.18	7.95	60.92
1 x 36W 1200mm, switchstart	nr	0.85	8.13	56.49	9.69	74.31
1 x 58W 1500mm, switchstart	nr	1.00	9.57	62.92	10.87	83.36
2 x 18W 600mm, switchstart	nr	0.75	7.18	57.70	9.73	74.61
2 x 36W 1200mm, switchstart	nr	1.00	9.57	65.73	11.29	75.30
2 x 58W 1500mm, switchstart	nr	1.25	11.96	80.70	13.90	106.56

M & E MEASUREMENT SERVICES

	Unit	Labour hours	Net labour (£)	Net material (£)	O'heads /profit (£)	Total (£)
Philips Pacific Emergency vandal proof luminaires (self-contained, maintained for 3 hours)						
1 x 36W, 1200mm	nr	0.95	9.09	156.49	24.84	165.58
1 x 58W, 1500mm	nr	1.10	10.53	162.92	26.02	173.45
2 x 36W, 1200mm	nr	1.10	10.53	165.73	26.44	202.70
2 x 58W, 1500mm	nr	1.35	12.92	180.70	29.04	222.66
Industrial discharge lighting						
Hermes 3 High Bay luminaires comprising gear, reflector and lamp						
SD 1570 - 150W SON lamp	nr	1.17	11.20	226.20	35.61	273.01
SD 2570 - 250W SON lamp	nr	1.17	11.20	245.76	38.54	295.50
SD 4570 - 400W SON lamp	nr	1.17	11.20	274.04	42.79	328.03
HD 2570 - 250W HPL/N lamp	nr	1.17	11.20	177.95	28.37	217.52
HD 2570 - 250W HPL/COM lamp	nr	1.17	11.20	178.24	28.42	217.86
HD 4070 - 400W HPL/N lamp	nr	1.17	11.20	197.48	31.30	239.98
HD 4070 - 400W HPL/COM lamp	nr	1.17	11.20	197.27	31.27	239.74
HD 4070 - 400W HPL/BUS lamp	nr	1.17	11.20	307.37	47.79	366.36
Low Bay Industrial Luminaires complete unit including						
150W SON/T lamp and gear	nr	1.17	11.20	90.55	15.26	117.01
250W SON/T lamp and gear	nr	1.17	11.20	109.20	18.06	138.46
400W SON/T lamp and gear	nr	1.17	11.20	120.31	19.73	151.24
250W SON/T comfort lamp and gear	nr	1.17	11.20	104.56	17.36	133.12
250W HPI/T lamp and gear	nr	1.17	11.20	143.13	23.15	177.48
400W HPI/T lamp and gear	nr	1.17	11.20	153.14	24.65	188.99
250W HPL comfort lamp and gear	nr	1.17	11.20	95.63	16.02	122.85
400W HPL comfort lamp and gear	nr	1.17	11.20	108.06	17.89	137.15
Covers, fitted						
wireguard	nr	0.25	2.39	16.49	2.83	21.71
plastic cover	nr	0.25	2.39	34.64	5.55	42.58
aluminium louvre	nr	0.25	2.39	22.10	3.67	28.16
glass cover	nr	0.25	2.39	32.69	5.26	40.34

LUMINAIRES AND LAMPS

	Unit	Labour hours	Net labour (£)	Net material (£)	O'heads /profit (£)	Total (£)
Road lighting						
Lanterns for SOX or SOXE lamps (includes for lantern, and gear only, not post or photocell or lamp)						
Group A for motorways and major road schemes						
90W - SOX - E66 lamp	nr	1.70	16.27	290.74	46.05	353.06
135W - SOX - E91 lamp	nr	1.70	16.27	390.77	61.06	468.10
180W - SOX - E131 lamp	nr	1.70	16.27	436.62	67.93	520.82
Group B road lighting including residential, commercial and industrial and amenity security lighting						
35W - SOX - E26 lamp	nr	1.70	16.27	170.95	28.08	215.30
55W - SOX - E36 lamp	nr	1.70	16.27	206.25	33.38	255.90
External security lantern/ bulkhead wallmounted, complete with lamps						
FGC100 bulkhead luminaire with						
1 x 9W PL lamp	nr	1.00	9.57	36.72	6.94	53.23
1 x 11W PL lamp	nr	1.00	9.57	41.99	7.73	59.29
Heavy duty bulkhead luminaires wire guard						
80W HPL-N lamps	nr	1.00	9.57	131.60	21.18	162.35
80W HPL-N lamps (zone 2)	nr	1.00	9.57	142.47	22.81	174.85
glass diffuser	nr	0.10	0.96	56.06	8.55	65.57
wire guard	nr	0.10	0.96	21.22	3.33	25.51
Philips HNF floodlight projectors						
Series 001 complete with lamp						
narrow beam 1kW HPI/T lamp	nr	1.60	15.31	443.88	68.88	528.07

M & E MEASUREMENT SERVICES

	Unit	Labour hours	Net labour (£)	Net material (£)	O'heads /profit (£)	Total (£)
narrow beam 2 x 400W HPI/T lamp	nr	1.60	15.31	451.72	70.05	537.08
wide beam 2 x 400W HPI/T lamp	nr	1.60	15.31	451.72	70.05	537.08
louvre	nr	0.20	1.91	114.43	17.45	133.79
square cornered front glass	nr	0.20	1.91	64.65	9.98	76.54
Series 002 complete with lamp						
narrow beam 2kW HPI/T lamp	nr	1.60	15.31	530.75	81.91	627.97
wide beam 2kW HPI/T lamp	nr	1.60	15.31	530.75	81.91	627.97
louvre	nr	0.20	1.91	114.43	17.45	133.79
square cornered front glass	nr	0.20	1.91	64.65	9.98	76.54
QVF enclosed floodlights (lamps not included)						
500W T/H lamp	nr	1.60	15.31	14.99	4.54	34.84
1000W T/H lamp	nr	1.60	15.31	19.99	5.29	40.59
1500W T/H lamp	nr	1.60	15.31	23.99	5.89	45.19

Downlights

Fully recessed downlighters excluding lamps and transformers, Philips model no.

	Unit	Labour hours	Net labour (£)	Net material (£)	O'heads /profit (£)	Total (£)
502W white for 12V 50W dichromic lamp	nr	0.70	6.70	8.00	2.21	14.70
502CH chrome for 12V 50W dichromic lamp	nr	0.70	6.70	10.71	2.61	17.41
522W for 100W halogen lamp	nr	0.70	6.70	38.90	6.84	52.44
528W for 2 x PLC 10 lamps	nr	0.70	6.70	64.18	10.63	81.51
590W for 2 x PLC 13 lamps	nr	0.70	6.70	89.36	14.41	110.47
602W white for 12V 50W halogen reflector lamp	nr	0.70	6.70	9.54	2.44	16.24
602G gold for 12V 50W halogen reflector lamp	nr	0.70	6.70	11.18	2.68	17.88
606W2 white for 60W lamp	nr	0.70	6.70	4.35	1.66	12.71
606G2 gold for 60W lamp	nr	0.70	6.70	5.80	1.88	14.38
608W2 white for 100W lamp	nr	0.70	6.70	4.35	1.66	12.71
608G2 gold for 100W lamp	nr	0.70	6.70	6.79	2.02	13.49
612W2 white for PAR38E lamp	nr	0.70	6.70	12.25	2.84	21.79
612G2 gold for PAR38E lamp	nr	0.70	6.70	14.40	3.17	24.27

LUMINAIRES AND LAMPS

	Unit	Labour hours	Net labour (£)	Net material (£)	O'heads /profit (£)	Total (£)
Recessed adjustable down lights, Philips model no.						
652W, white for 12V 50W dichromic lamp	nr	0.70	6.70	25.55	4.84	32.25
652CH, chrome for 12V 50W dichroic lamp	nr	0.70	6.70	35.03	6.26	47.99
652G, gold for 12V 50W dichromic lamp	nr	0.70	6.70	38.95	6.85	45.65
656W, white for R63 lamp	nr	0.70	6.70	12.23	2.84	21.77
656G, gold for R63 lamp	nr	0.70	6.70	18.34	3.76	28.80
Low voltage capsule downlighters, Philips model no.						
QBG902W, white for 12V 10W halogen capsule	nr	0.70	6.70	15.63	3.35	25.68
QBG902CH, chrome for 12V 10W halogen capsule	nr	0.70	6.70	18.07	3.72	28.49
QBG902G, gold for 12V 10W halogen capsule	nr	0.70	6.70	18.07	3.72	28.49
OBG903W, white for 12V 20W halogen capsule	nr	0.70	6.70	25.50	4.83	32.20
OBG903CH, chrome for 12V 20W halogen capsule	nr	0.70	6.70	27.80	5.17	34.50
OBG903G, gold for 12V 20W halogen capsule	nr	0.70	6.70	27.80	5.17	34.50
Spotlight projectors, complete with electronic transformer and dichromic lamp, excluding track						
Low voltage Litapacks, Philips model no						
502 WLP white	nr	0.50	4.79	35.01	5.97	45.77
502 WGP gold	nr	0.50	4.79	36.57	6.20	47.56
502 CHLP chrome	nr	0.50	4.79	36.57	7.28	41.36
641 WLP white	nr	0.50	4.79	51.83	8.49	65.11
641 GLP gold	nr	0.50	4.79	58.99	9.57	73.35
641 CHLP chrome	nr	0.50	4.79	58.99	9.57	73.35
651 WLP white	nr	0.50	4.79	43.41	8.31	48.20

M & E MEASUREMENT SERVICES

	Unit	Labour hours	Net labour (£)	Net material (£)	O'heads /profit (£)	Total (£)
652 WLP white	nr	0.50	4.79	51.20	8.40	64.39
652 GLP gold	nr	0.50	4.79	60.10	9.73	64.89
652 CHLP chrome	nr	0.50	4.79	60.10	9.73	74.62

Mid-series conical range, for
direct contact system

	Unit	Labour hours	Net labour (£)	Net material (£)	O'heads /profit (£)	Total (£)
white for ES PAR 80/120W	nr	0.50	4.79	27.35	4.82	36.96
white for ES R63 40/60W	nr	0.50	4.79	31.90	5.50	42.19
white for ES R80 60/75/100W	nr	0.50	4.79	36.70	6.22	47.71
white for ES 60W bowl reflector	nr	0.50	4.79	36.70	6.22	47.71
white for ES 100W bowl reflector	nr	0.50	4.79	40.45	6.79	52.03
white for 12V 50W dichromic lamp including transformer	nr	0.70	6.70	68.35	11.26	86.31

Spotlight track systems

Lita twin circuit track modules

	Unit	Labour hours	Net labour (£)	Net material (£)	O'heads /profit (£)	Total (£)
1.5m track length, aluminium	nr	0.60	5.74	19.61	3.80	25.35
2.5m track length, aluminium	nr	0.90	8.61	32.33	6.14	40.94
3.5m track length, aluminium	nr	1.20	11.48	45.37	8.53	56.85
2.5m recessed track length, aluminium	nr	0.90	8.61	36.15	6.50	44.76
3.5m recessed track length, aluminium	nr	1.20	11.48	49.55	9.15	61.03
3.5m track length, black	nr	1.20	11.48	49.82	9.20	61.30
1.5m track length, white	nr	0.60	5.74	22.63	4.26	28.37
2.5m track length, white	nr	0.90	8.61	37.26	6.88	45.87
3.5m track length, white	nr	1.20	11.48	52.20	9.55	63.68

Lita twin circuit track supply
and connection accessories

	Unit	Labour hours	Net labour (£)	Net material (£)	O'heads /profit (£)	Total (£)
dead end, white	nr	0.10	0.96	1.20	0.32	2.16
straight connector	nr	0.17	1.63	3.71	0.80	6.14
universal supply box including live end and dead end, white	nr	0.33	3.16	8.37	1.73	11.53
flexible coupler	nr	0.33	3.16	19.29	3.37	25.82
L, T or X coupler, cube	nr	0.17	1.63	4.88	0.98	6.51
hook adaptor	nr	0.17	1.63	10.18	1.77	13.58

LUMINAIRES AND LAMPS

	Unit	Labour hours	Net labour (£)	Net material (£)	O'heads /profit (£)	Total (£)
Montana II LV minitrack modules						
1m length track, white	nr	0.50	4.79	14.79	2.94	22.52
1m length track, black	nr	0.50	4.79	14.79	2.94	22.52
2m length track, white	nr	0.70	6.70	22.10	4.32	33.12
2m length track, black	nr	0.70	6.70	22.10	4.32	33.12
3m length track, white	nr	0.90	8.61	28.12	5.51	42.24
3m length track, black	nr	0.90	8.61	28.12	5.51	42.24
Montana II LV minitrack supply and connection accessories						
end power supply	nr	0.40	3.83	4.61	1.27	9.71
electrical in-line coupler	nr	0.36	3.45	4.61	1.21	8.06
dead end cap	nr	0.10	0.96	1.70	0.40	3.06
wire suspension set	nr	1.00	9.57	3.42	1.95	14.94
cover strip	nr	0.10	0.96	4.61	0.84	6.41
90 degrees fixed corner	nr	0.30	2.87	9.27	1.82	12.14
adaptor for 'Jack' connection projectors	nr	0.17	1.63	3.13	0.71	5.47

Lamps, fitted to luminaires

GLS lamps, bayonet cap connection, running on 240W

	Unit	Labour hours	Net labour (£)	Net material (£)	O'heads /profit (£)	Total (£)
25W, pearl	nr	0.10	0.96	0.87	0.27	1.83
25W, clear	nr	0.10	0.96	0.89	0.28	2.13
40W, pearl	nr	0.10	0.96	0.69	0.25	1.90
40W, clear	nr	0.10	0.96	0.80	0.26	2.02
60W, pearl	nr	0.10	0.96	0.69	0.25	1.90
60W, clear	nr	0.10	0.96	0.80	0.26	2.02
75W, pearl	nr	0.10	0.96	0.93	0.28	2.17
100W, pearl	nr	0.10	0.96	0.69	0.25	1.65
100W, clear	nr	0.10	0.96	0.80	0.26	2.02
150W, pearl	nr	0.10	0.96	1.18	0.32	2.46
150W, clear	nr	0.10	0.96	1.22	0.33	2.51

M & E MEASUREMENT SERVICES

	Unit	Labour hours	Net labour (£)	Net material (£)	O'heads /profit (£)	Total (£)
Spotline blown bulb reflector/ internally silvered, 240W with screw or bayonet cup connection, type						
R63 40W	nr	0.10	0.96	2.06	0.45	3.47
R63 60W	nr	0.10	0.96	2.06	0.45	3.47
R80 60W	nr	0.10	0.96	2.06	0.45	3.47
R80 75W	nr	0.10	0.96	2.06	0.45	3.47
R80 100W	nr	0.10	0.96	2.06	0.45	3.47
R95 75W	nr	0.10	0.96	3.34	0.64	4.94
R95 100W	nr	0.10	0.96	3.34	0.64	4.94
R95 150W	nr	0.10	0.96	4.90	0.88	6.74
Tungsten halogen display lamps						
Standard dichromic reflector lamps open front, operating on 1VV with 2-pin base, Philips code						
M68 20W 12 degrees	nr	0.10	0.96	8.87	1.47	9.83
M69 20W 36 degrees	nr	0.10	0.96	8.87	1.47	9.83
M81 35W 36 degrees	nr	0.10	0.96	8.87	1.47	11.30
M49 50W 12 degrees	nr	0.10	0.96	8.87	1.47	11.30
M50 50W 24 degrees	nr	0.10	0.96	8.87	1.47	11.30
M58 50W 36 degrees	nr	0.10	0.96	8.87	1.47	11.30
Tungsten halogen display capsules (12V)						
M90 5W	nr	0.10	0.96	5.95	1.04	6.91
M91 10W	nr	0.10	0.96	5.95	1.04	7.95
M88 20W	nr	0.10	0.96	5.95	1.04	7.95
M89 50W	nr	0.10	0.96	5.37	0.95	7.28
M90 100W	nr	0.10	0.96	7.61	1.29	9.86
Tungsten halogen floodlight lamps operating on mains voltage						
100T 3Q/CL/P 100W	nr	0.15	1.44	7.89	1.40	10.73
150T 3Q/CL/P 150W	nr	0.15	1.44	7.89	1.40	10.73
200T 3Q/CL/P 200W	nr	0.15	1.44	7.43	1.33	10.20
300T 3Q/CL/P 300W	nr	0.15	1.44	7.98	1.41	10.83
500T 3Q/CL/P 500W	nr	0.15	1.44	12.85	2.14	14.29

Tungsten floodlights (cont'd)	Unit	Labour hours	Net labour (£)	Net material (£)	O'heads /profit (£)	Total (£)
1000T 3Q/CL/P 1000W	nr	0.15	1.44	13.78	2.28	15.22
1500T 3Q/CL/P 1500W	nr	0.15	1.44	15.12	2.48	16.56
2000T 3Q/CL/P 2000W	nr	0.15	1.44	17.11	2.78	21.33

Compact fluorescent lamps running on mains voltage

PLS lamps 2 pins with integral starter

	Unit	Labour hours	Net labour (£)	Net material (£)	O'heads /profit (£)	Total (£)
5W	nr	0.10	0.96	4.37	0.80	6.13
7W	nr	0.10	0.96	4.37	0.80	6.13
9W	nr	0.10	0.96	4.37	0.80	6.13
11W	nr	0.10	0.96	4.37	0.80	6.13

PLS lamps 4 pins without integral starter

	Unit	Labour hours	Net labour (£)	Net material (£)	O'heads /profit (£)	Total (£)
5W	nr	0.10	0.96	5.91	1.03	7.90
7W	nr	0.10	0.96	5.91	1.03	7.90
9W	nr	0.10	0.96	5.91	1.03	7.90
11W	nr	0.10	0.96	5.91	1.03	6.87

PLL lamps without integral starter

	Unit	Labour hours	Net labour (£)	Net material (£)	O'heads /profit (£)	Total (£)
18W	nr	0.10	0.96	8.53	1.42	10.91
24W	nr	0.10	0.96	8.53	1.42	10.91
36W	nr	0.10	0.96	9.75	1.61	12.32
40W	nr	0.10	0.96	10.18	1.67	12.81
50W	nr	0.10	0.96	10.72	1.75	11.68

Straight fluorescent lamps standard white 26mm TLD tubes

	Unit	Labour hours	Net labour (£)	Net material (£)	O'heads /profit (£)	Total (£)
450mm 15W	nr	0.10	0.96	7.88	1.33	10.17
600mm 18W	nr	0.10	0.96	6.90	1.18	9.04
900mm 30W	nr	0.10	0.96	7.80	1.31	10.07
1200mm 36W	nr	0.10	0.96	8.72	1.45	11.13
1500mm 58W	nr	0.12	1.15	9.36	1.58	12.09
1800mm 70W	nr	0.13	1.24	11.08	1.85	14.17
2400mm 1000W	nr	0.15	1.44	15.88	2.60	19.92

M & E MEASUREMENT SERVICES

	Unit	Labour hours	Net labour (£)	Net material (£)	O'heads /profit (£)	Total (£)
Low pressure sodium lamps, sox						
35W	nr	0.15	1.44	23.55	3.75	28.74
55W	nr	0.15	1.44	27.48	4.34	33.26
90W	nr	0.15	1.44	33.36	5.22	40.02
135W	nr	0.15	1.44	43.17	6.69	51.30
180W	nr	0.15	1.44	68.68	10.52	80.64
High pressure sodium lamps, standard						
50W	nr	0.15	1.44	36.35	5.67	43.46
70W	nr	0.15	1.44	39.60	6.16	47.20
150W	nr	0.15	1.44	48.24	7.45	57.13
250W	nr	0.15	1.44	69.95	10.71	82.10
400W	nr	0.17	1.63	84.80	12.96	99.39
1000W	nr	0.20	1.91	183.36	27.79	213.06
Mercury fluorescent lamps						
50W	nr	0.15	1.44	12.88	2.15	16.47
80W	nr	0.15	1.44	12.88	2.15	16.47
125W	nr	0.15	1.44	15.36	2.52	19.32
250W	nr	0.15	1.44	28.92	4.55	34.91
400W	nr	0.17	1.63	44.51	6.92	53.06
700W	nr	0.17	1.63	81.53	12.47	95.63
1000W	nr	0.19	1.82	106.29	16.22	108.11
Metal halide						
250W	nr	0.15	1.44	67.41	10.33	79.18
400W	nr	0.15	1.44	90.28	13.76	105.48
1000W	nr	0.19	1.82	244.86	37.00	246.68
2000W	nr	0.21	2.01	371.54	56.03	429.58

ACCESSORIES FOR ELECTRIC SERVICES

	Unit	Labour hours	Net labour (£)	Net material (£)	O'heads /profit (£)	Total (£)
Y74 ACCESSORIES FOR ELECTRICAL SERVICES						

General notes.

1. All times allow for cable connections to wiring accessories

2. Material costs are based on split pack prices

3. A discount of 24% has been incorporated within the net material costs in this section, excluding chimes and controls

Fittings (MK Ltd)

Plate switches: 6 amp flush mounted, white plastic, fixed to backgrounds

	Unit	Labour hours	Net labour (£)	Net material (£)	O'heads /profit (£)	Total (£)
1 gang, 1 way SP	nr	0.17	1.63	1.31	0.44	3.38
1 gang, 1 way DP	nr	0.19	1.82	3.27	0.76	5.85
1 gang, 2 way SP	nr	0.17	1.63	1.55	0.48	3.66
2 gang, 2 way SP	nr	0.25	2.39	2.80	0.78	5.97
3 gang, 2 way SP	nr	0.38	3.64	4.02	1.15	8.81
4 gang, 2 way SP	nr	0.43	4.12	6.88	1.65	12.65
1 gang, intermediate	nr	0.17	1.63	3.72	0.80	5.35
1 gang, push switch SP	nr	0.17	1.63	2.33	0.59	4.55

Architrave plate switches: 6 amp flush mounted, white plastic, fixed to backgrounds with screws

	Unit	Labour hours	Net labour (£)	Net material (£)	O'heads /profit (£)	Total (£)
1 gang, 2 way SP	nr	0.17	1.63	1.62	0.49	3.74
2 gang, 2 way SP	nr	0.25	2.39	3.42	0.87	6.68
1 gang, push switch SP	nr	0.17	1.63	2.52	0.62	4.77

M & E MEASUREMENT SERVICES

	Unit	Labour hours	Net labour (£)	Net material (£)	O'heads /profit (£)	Total (£)
Dimmer switches: 5 amp flush mounted, white plastic, fixed to backgrounds with screws						
1 gang 400W with 5 amp, 2 way switch	nr	0.25	2.39	17.52	2.99	22.90
1 gang 400W, rotary action	nr	0.17	1.63	11.62	1.99	15.24
2 gang 250W, rotary action	nr	0.25	2.39	22.37	3.71	28.47
DP switches: white plastic, flush mounted, fixed to backgrounds with screws						
20 amp, 1 gang	nr	0.28	2.68	3.72	0.96	7.36
20 amp, 1 gang with neon	nr	0.28	2.68	5.34	1.20	9.22
20 amp, 1 gang with neon, and mounted water heater	nr	0.42	4.02	6.08	1.51	11.61
20 amp, 1 gang with flex outlet	nr	0.42	4.02	3.72	1.16	8.90
20 amp, 1 gang with flex outlet and neon	nr	0.42	4.02	5.77	1.47	11.26
32 amp, 1 gang with neon	nr	0.28	2.68	6.48	1.37	10.53
45 amp, 1 gang moulded	nr	0.32	3.06	4.97	1.20	9.23
45 amp, 1 gang moulded with neon	nr	0.32	3.06	5.78	1.33	10.17
DP switches: surface mounted metal fixed to backgrounds with screws						
63 amp, 1 gang with neon	nr	0.32	3.06	19.28	3.35	25.69
16kW DP main switch with 20 amp DP switch and red neon indicator	nr	0.42	4.02	25.43	4.42	29.45
TP&N, 32 amp, metal switch fixed to backgrounds with screws						
flush mounted with neon	nr	0.42	4.02	14.87	2.83	21.72
surface mounted with neon	nr	0.42	4.02	16.75	3.12	23.89
Socket outlets: 13 amp flush mounted, white plastic, fixed to backgrounds with screws						
1 gang	nr	0.23	2.20	2.28	0.67	5.15
2 gang	nr	0.28	2.68	4.08	1.01	7.77
1 gang switched	nr	0.23	2.20	2.91	0.77	5.11

ACCESSORIES FOR ELECTRIC SERVICES

Socket outlets (cont'd)	Unit	Labour hours	Net labour (£)	Net material (£)	O'heads /profit (£)	Total (£)
2 gang switched	nr	0.28	2.68	5.64	1.25	8.32
1 gang switched with 1 neon	nr	0.23	2.20	5.22	1.11	8.53
2 gang switched with 1 neon	nr	0.28	2.68	9.31	1.80	13.79
2 gang switched filtered socket	nr	0.28	2.68	19.72	3.36	25.76
RCD protected socket outlet: 13 amp, flush mounted, white plastic, fixed to backgrounds with screws						
10mA active control circuit	nr	0.28	2.68	33.17	5.38	41.23
30mA active control ciruit	nr	0.28	2.68	31.90	5.19	39.77
30mA passive control ciruit	nr	0.28	2.68	31.90	5.19	39.77
Fused connection units: 13 amp flush mounted, white plastic fixed to backgrounds with screws						
unswitched	nr	0.42	4.02	4.39	1.26	9.67
unswitched with flex outlet	nr	0.42	4.02	4.53	1.28	9.83
unswitched with neon	nr	0.42	4.02	6.31	1.55	11.88
unswitched with flex outlet and neon	nr	0.42	4.02	7.01	1.65	12.68
DP switched	nr	0.42	4.02	4.92	1.34	10.28
DP switched with flex outlet	nr	0.42	4.02	5.11	1.37	10.50
DP switched with neon	nr	0.42	4.02	6.79	1.62	12.43
DP switched with flex outlet and neon	nr	0.42	4.02	7.05	1.66	12.73
Shaver outlets: flush mounted, white plastic fixed to backgrounds with screws						
fused shaver socket outlet	nr	0.33	3.16	10.07	1.98	15.21
dual voltage shaver supply unit	nr	0.33	3.16	19.93	3.46	26.55
Blank plates: white plastic, fixed to backgrounds with screws						
1 gang, moulded	nr	0.10	0.96	0.82	0.27	1.78
2 gang, moulded	nr	0.10	0.96	1.76	0.41	2.72
1 gang, moulded architrave	nr	0.10	0.96	1.30	0.34	2.26

M & E MEASUREMENT SERVICES

	Unit	Labour hours	Net labour (£)	Net material (£)	O'heads /profit (£)	Total (£)
Cooker connection/control units, white plastic, fixing with screws to background						
flush connection unit	nr	0.38	3.64	3.19	1.02	7.85
flush, 45 amp DP main switch and 13 amp switch socket outlet	nr.	0.42	4.02	9.49	2.03	15.54
flush, 45 amp DP main switch and 13 amp switch socket outlet with neon	nr	0.42	4.02	12.45	2.47	18.94
surface, 45 amp DP main switch and 13 amp switch outlet	nr	0.42	4.02	9.55	2.04	15.61
surface, 45 amp DP main switch and 13 amp switch socket outlet with neon	nr	0.42	4.02	12.55	2.49	19.06
Cooker control units, metal fixing with screws to backgrounds						
flush 45 amp DP main switch and 13 amp switch socket outlet with 1 neon	nr	0.42	4.02	12.49	2.48	18.99
flush, 45 amp DP switch with neon	nr	0.38	3.64	9.33	1.95	14.92
surface, 45 amp DP main switch and 13 amp switch socket outlet with neon	nr	0.42	4.02	16.90	3.14	20.92
Metal finish fittings, finishes golden bronze, stainless steel, stainless steel highly polished, stainless steel brushed, flush mounted, fixed to backgrounds with screws						
6 amp plate switches						
1 gang 2 way	nr	0.17	1.63	4.29	0.89	6.81
2 gang 2 way	nr	0.25	2.39	6.02	1.26	9.67
3 gang 2 way	nr	0.38	3.64	9.19	1.92	14.75
1 gang intermediate	nr	0.17	1.63	7.94	1.44	11.01
bell push	nr	0.17	1.63	6.00	1.14	8.77

ACCESSORIES FOR ELECTRIC SERVICES

	Unit	Labour hours	Net labour (£)	Net material (£)	O'heads /profit (£)	Total (£)
13 amp socket outlets						
1 gang unswitched	nr	0.23	2.20	5.08	1.09	8.37
2 gang unswitched	nr	0.28	2.68	9.29	1.80	13.77
1 gang switched	nr	0.23	2.20	5.62	1.17	8.99
2 gang switched	nr	0.28	2.68	10.15	1.92	14.75
1 gang switched with neon	nr	0.23	2.20	8.34	1.58	12.12
2 gang switched with neon	nr	0.28	2.68	15.26	2.69	20.63
13 amp fused connection units						
switched	nr	0.42	4.02	7.42	1.72	13.16
unswitched	nr	0.42	4.02	7.01	1.65	12.68
switched with cable outlet	nr	0.42	4.02	7.93	1.79	13.74
switched with neon	nr	0.42	4.02	10.19	2.13	16.34
unswitched with cable outlet	nr	0.42	4.02	7.01	1.65	12.68
switched with neon and cable outlet	nr	0.42	4.02	10.62	2.20	16.84
DP switches						
20 amp, with neon	nr	0.42	4.02	8.44	1.87	14.33
20 amp, flex outlet and neon	nr	0.42	4.02	8.12	1.82	13.96
32 amp, with neon	nr	0.42	4.02	10.65	2.20	16.87
45 amp, with neon	nr	0.42	4.02	10.03	2.11	14.05
TP&N 32 amp switch with neon	nr	0.42	4.02	21.46	3.82	29.30
dual voltage shaver supply unit	nr	0.33	3.16	23.07	3.93	26.23
Blank plates						
1 gang	nr	0.10	0.96	2.93	0.58	4.47
2 gang	nr	0.10	0.96	5.72	1.00	7.68
Steel boxes, fixed to backgrounds requiring drilling, plugging and screwing						
Flush mounted						
16mm deep, 1 gang	nr	0.25	2.39	0.75	0.47	3.61
25mm deep, 1 gang	nr	0.25	2.39	0.74	0.47	3.60
25mm deep, 2 gang	nr	0.25	2.39	0.96	0.50	3.85
35mm deep, 1 gang	nr	0.25	2.39	0.85	0.49	3.73
35mm deep, 2 gang	nr	0.25	2.39	1.11	0.53	4.03

	Unit	Labour hours	Net labour (£)	Net material (£)	O'heads /profit (£)	Total (£)
35mm deep, dual box	nr	0.25	2.39	1.46	0.58	4.43
46mm deep, 1 gang	nr	0.28	2.68	1.21	0.58	4.47
47mm deep, 2 gang	nr	0.28	2.68	1.89	0.69	5.26
for cooker control unit 152 x						
140 x 55mm	nr	0.42	4.02	3.08	1.06	8.16
for 32 amp TP&N switches 178 x						
114 x 65mm	nr	0.42	4.02	4.12	1.22	9.36
architrave, 2 gang	nr	0.25	2.39	1.07	0.52	3.98
Surface mounted						
40mm deep, 1 gang	nr	0.28	2.68	1.63	0.65	4.96
40mm deep, 2 gang	nr	0.28	2.68	2.34	0.75	5.77

White plastic boxes, fixed to backgrounds requiring drilling, plugging and screwing

Flush flange boxes

	Unit	Labour hours	Net labour (£)	Net material (£)	O'heads /profit (£)	Total (£)
41mm deep, 1 gang	nr	0.28	2.68	0.96	0.55	4.19
41mm deep, 2 gang	nr	0.28	2.68	1.33	0.60	4.01

Surface mounted

	Unit	Labour hours	Net labour (£)	Net material (£)	O'heads /profit (£)	Total (£)
19mm deep, 1 gang	nr	0.25	2.39	0.83	0.48	3.70
19mm deep, 2 gang	nr	0.25	2.39	2.17	0.68	5.24
32mm deep, 1 gang	nr	0.25	2.39	0.87	0.49	3.75
32mm deep, 2 gang	nr	0.25	2.39	1.50	0.58	4.47
44mm deep, 1 gang	nr	0.28	2.68	1.01	0.55	4.24
44mm deep, 2 gang	nr	0.28	2.68	2.57	0.79	6.04
38mm deep, dual box	nr	0.35	3.35	1.94	0.79	6.08
architrave, 1 gang	nr	0.25	2.39	0.96	0.50	3.85
architrave, 2 gang	nr	0.25	2.39	2.26	0.70	5.35

Metal clad fittings; including back box, surface mounted, fixed to backgrounds requiring drilling, plugging and screwing

6 amp switches

	Unit	Labour hours	Net labour (£)	Net material (£)	O'heads /profit (£)	Total (£)
1 gang, 1 way	nr	0.42	4.02	3.90	1.19	9.11
2 gang, 1 way	nr	0.50	4.79	5.11	1.49	11.39

ACCESSORIES FOR ELECTRIC SERVICES

Metal clad fittings (cont'd)	Unit	Labour hours	Net labour (£)	Net material (£)	O'heads /profit (£)	Total (£)
DP switches						
20 amp, 1 gang	nr	0.53	5.07	5.34	1.56	11.97
20 amp, 1 gang with neon	nr	0.53	5.07	7.24	1.85	14.16
20 amp, 1 gang with neon and flex outlet	nr	0.67	6.41	7.55	2.09	16.05
45 amp, 1 gang with neon	nr	0.53	5.07	9.70	2.22	16.99
32 amp TP&N switch with neon	nr	0.67	6.41	17.62	3.60	27.63
13 amp socket outlets						
1 gang unswitched	nr	0.48	4.59	3.90	1.27	9.76
2 gang unswitched	nr	0.53	5.07	7.22	1.84	14.13
1 gang switched	nr	0.48	4.59	5.42	1.50	11.51
2 gang switched	nr	0.53	5.07	10.64	2.36	18.07
1 gang switched with neon	nr	0.48	4.59	7.76	1.85	14.20
2 gang switched with neon	nr	0.53	5.07	14.11	2.88	22.06
13 amp, fused connection units						
unswitched	nr	0.67	6.41	5.24	1.75	13.40
unswitched with flex outlet	nr	0.67	6.41	5.67	1.81	13.89
DP switched	nr	0.67	6.41	6.07	1.87	14.35
DP switched with neon	nr	0.67	6.41	7.27	2.05	15.73
DP switched with flex outlet	nr	0.67	6.41	6.55	1.94	14.90
DP switched with flex outlet and neon	nr	0.67	6.41	7.58	2.10	16.09
Gridswitch system						
Gridswitches, clipped direct to grids						
6 amp 1 way SP	nr	0.35	3.35	1.27	0.69	5.31
6 amp 2 way SP	nr	0.38	3.64	1.68	0.80	6.12
20 amp 1 way SP	nr	0.35	3.35	1.76	0.84	5.11
20 amp 2 way SP	nr	0.35	3.35	2.26	0.84	6.45
20 amp intermediate	nr	0.40	3.83	2.56	0.96	7.35
20 amp 1 way DP	nr	0.40	3.83	2.87	1.00	7.70
20 amp, 2 way SP key switch	nr	0.35	3.35	3.97	1.10	8.42
20 amp 1 way DP key switch	nr	0.40	3.83	3.87	1.16	8.86
20 amp intermediate key switch	nr	0.40	3.83	4.43	1.24	9.50
6 amp 2 way push switch	nr	0.35	3.35	3.01	0.81	6.36

M & E MEASUREMENT SERVICES

	Unit	Labour hours	Net labour (£)	Net material (£)	O'heads /profit (£)	Total (£)
6 amp 2 way push switch with						
bell symbol	nr	0.35	3.35	3.16	0.98	7.49
blank insert	nr	0.02	0.19	0.81	0.15	1.15
neon indicator (red)	nr	0.30	2.87	4.31	1.08	7.18
Grids, fixed to backgrounds with screws						
1 gang	nr	0.17	1.63	0.75	0.36	2.74
2 gang	nr	0.17	1.63	0.75	0.36	2.74
3 and 4 gang	nr	0.20	1.91	0.97	0.43	3.31
6 gang for (18 and 24 gang)	nr	0.25	2.39	2.01	0.66	5.06
1 gang architrave	nr	0.17	1.63	0.82	0.37	2.82
Steel boxes, fixed to backgrounds requiring drilling, plugging and screwing						
1 and 2 gang	nr	0.13	1.24	0.93	0.33	2.50
3 and 4 gang	nr	0.23	2.20	1.55	0.56	4.31
6 and 8 gang	nr	0.23	2.20	2.46	0.70	5.36
9 and 12 gang	nr	0.25	2.39	2.93	0.80	6.12
18 gang	nr	0.30	2.87	6.86	1.46	11.19
24 gang	nr	0.33	3.16	8.33	1.72	13.21
1 gang architrave	nr	0.10	0.96	1.35	0.35	2.66
Cover plates fixed to backgrounds with screws						
Moulded flush plates						
1 gang	nr	0.15	1.44	0.75	0.33	2.52
2 gang	nr	0.25	2.39	0.74	0.47	3.60
3 gang	nr	0.25	2.39	1.56	0.59	4.54
4 gang	nr	0.25	2.39	1.56	0.59	4.54
6 gang	nr	0.25	2.39	2.23	0.69	5.31
8 gang	nr	0.25	2.39	2.23	0.69	5.31
Metal flush plates, matt chrome and satin brass						
1 gang	nr	0.15	1.44	3.06	0.67	5.17
2 gang	nr	0.25	2.39	3.06	0.82	6.27
3 gang	nr	0.25	2.39	5.14	1.13	8.66

Metal flush plates (cont'd)	Unit	Labour hours	Net labour (£)	Net material (£)	O'heads /profit (£)	Total (£)
4 gang	nr	0.25	2.39	5.14	1.13	8.66
6 gang	nr	0.25	2.39	8.10	1.57	12.06
8 gang	nr	0.25	2.39	8.10	1.57	12.06
9 gang	nr	0.35	3.35	10.40	2.06	15.81
12 gang	nr	0.35	3.35	10.61	2.09	16.05
18 gang	nr	0.40	3.83	16.33	3.02	23.18
24 gang	nr	0.42	4.02	21.43	3.82	29.27
1 gang architrave	nr	0.15	1.44	2.80	0.64	4.88

Matt chrome surface plates

	Unit	Labour hours	Net labour (£)	Net material (£)	O'heads /profit (£)	Total (£)
1 gang	nr	0.15	1.44	3.51	0.74	5.69
2 gang	nr	0.25	2.39	3.51	0.89	6.79
3 gang	nr	0.25	2.39	5.84	1.23	9.46
4 gang	nr	0.25	2.39	5.84	1.23	9.46
6 gang	nr	0.25	2.39	7.84	1.53	11.76
8 gang	nr	0.25	2.39	7.84	1.53	11.76
9 gang	nr	0.35	3.35	9.36	1.91	14.62
12 gang	nr	0.35	3.35	9.54	1.93	14.82
18 gang	nr	0.40.	3.83	14.66	2.77	21.26
24 gang	nr	0.42	4.02	19.27	3.49	26.78

Metalclad surface plates

	Unit	Labour hours	Net labour (£)	Net material (£)	O'heads /profit (£)	Total (£)
1 gang	nr	0.15	1.44	0.92	0.35	2.71
2 gang	nr	0.25	2.39	0.92	0.50	3.81
3 gang	nr	0.25	2.39	1.82	0.63	4.84
4 gang	nr	0.25	2.39	1.82	0.63	4.84
6 gang	nr	0.25	2.39	3.18	0.84	6.41
8 gang	nr	0.35	3.35	3.18	0.98	7.51
9 gang	nr	0.35	3.35	4.92	1.24	9.51
12 gang	nr	0.35	3.35	5.02	1.26	9.63
18 gang	nr	0.40	3.83	10.35	2.13	16.31
24 gang	nr	0.42	4.02	13.74	2.66	20.42

Lampholders, white plastic, fixed
to backgrounds with screws

	Unit	Labour hours	Net labour (£)	Net material (£)	O'heads /profit (£)	Total (£)
combined block/ceiling rose	nr	0.30	2.87	1.10	0.60	4.57
6" pendant set	nr	0.60	5.74	2.61	1.25	9.60
6" plug in pendant set	nr	0.60	5.74	5.27	1.65	12.66
heat resistant pendant	nr	0.30	2.87	1.28	0.62	4.77
batten 2T mounted in base with skirt	nr	0.30	2.87	2.82	0.85	6.54

M & E MEASUREMENT SERVICES

	Unit	Labour hours	Net labour (£)	Net material (£)	O'heads /profit (£)	Total (£)
batten 3T mounted in base with skirt	nr	0.30	2.87	3.15	0.90	6.92
angle battens 3T with skirt	nr	0.30	2.87	2.28	0.77	5.92
Junction boxes, fixed to backgrounds with screws						
20 amp, 4 terminal	nr	0.80	7.66	1.32	1.35	10.33
20 amp, 3 terminal	nr	0.80	7.66	1.63	1.39	10.68
Ceiling switches, fixed to backgrounds with screws						
6 amp, 1 way SP surface mounted	nr	0.30	2.87	3.03	0.89	6.79
6 amp, 2 way SP surface mounted	nr	0.30	2.87	3.47	0.95	7.29
45 amp, 1 way DP with pilot lamp flush mounted	nr	0.30	2.87	8.97	1.78	13.62
Plugs/adaptors to BS1363/A1994						
fitted with 13 amp fuse	nr	0.20	1.91	1.38	0.49	3.78
fitted with 3 amp fuse	nr	0.20	1.91	1.36	0.49	3.76

V & E Friedland Ltd. A discount of 10% has been incorporated within the nett material costs in this section

Chimes fixed to backgrounds requiring drilling, plugging and screwing including connection of bell line

	Unit	Labour hours	Net labour (£)	Net material (£)	O'heads /profit (£)	Total (£)
Gala	nr	0.60	5.74	5.52	1.69	12.95
Facet	nr	0.60	5.74	6.61	1.85	14.20
Cameo	nr	6.89	65.94	6.89	10.92	83.75
Ding Dong	nr	0.60	5.74	7.27	1.95	14.96
Big Ben	nr	0.60	5.74	7.76	2.02	15.52
Stratum	nr	0.60	5.74	7.67	2.01	15.42
Gemini	nr	0.60	5.74	9.50	2.29	17.53
Orbit	nr	0.60	5.74	10.57	2.45	18.76
Quartet	nr	0.60	5.74	11.35	2.56	19.65
Regent	nr	0.60	5.74	10.63	2.46	18.83
Duet	nr	0.60	5.74	14.54	3.04	23.32

ACCESSORIES FOR ELECTRIC SERVICES

Chimes (cont'd)	Unit	Labour hours	Net labour (£)	Net material (£)	O'heads /profit (£)	Total (£)
Sahara	nr	0.60	5.74	12.12	2.68	20.54
Rialto	nr	0.60	5.74	13.79	2.93	22.46
Vouge	nr	0.60	5.74	15.48	3.18	24.40
Warbler	nr	0.60	5.74	18.38	3.62	27.74
York	nr	0.60	5.74	19.62	3.80	29.16
Bells and buzzers						
Underdome	nr	0.50	4.79	4.00	1.32	10.11
Bell-in-one	nr	0.50	4.79	6.09	1.63	12.51
Minibuzzer	nr	0.50	4.79	2.93	1.16	8.88
Pushes						
Dimex	nr	0.50	4.79	1.04	0.16	5.83
Sesame	nr	0.50	4.79	1.23	0.90	6.92
Gem	nr	0.50	4.79	2.67	1.12	8.58
Button	nr	0.50	4.79	1.04	0.87	6.70
Light spot (white)	nr	0.50	4.79	1.81	0.99	7.59
Lighted button	nr	0.50	4.79	2.44	1.08	8.31
Transformers						
240 to 8V	nr	0.60	5.74	10.11	2.38	15.85
240 to 12V	nr	0.60	5.74	12.81	2.78	18.55

Superswitch

Time controls wired-in fixing
to backgrounds requiring drilling
plugging and screwing

24 hour immersion heater timer	nr	1.00	9.57	25.45	5.25	40.27
7 day immersion heater timer	nr	1.00	9.57	36.20	6.87	52.64
24 hour central heating programmer	nr	1.00	9.57	35.71	6.79	52.07
7 day central heating programmer	nr	1.00	9.57	39.00	7.29	55.86
24 hour multi-purpose timer	nr	1.00	9.57	25.45	5.25	40.27
7 day multi-purpose timer	nr	1.00	9.57	35.00	6.69	51.26

	Unit	Labour hours	Net labour (£)	Net material (£)	O'heads /profit (£)	Total (£)

Y81 TESTING AND COMMISSIONING OF ELECTRICAL SERVICES

IEE Regulations requires the testing of all electrical installations. The tests given below are for a typical house installation and the figures represent the costs of testing each individual circuit. The costs of remedial work are excluded.

Waiting time for meeting an Inspector/Clerk of Works to verify the tests is also excluded.

	Unit	Labour hours	Net labour (£)	Net material (£)	O'heads /profit (£)	Total (£)
Continuity of ring final circuit conductors	nr	0.75	7.18	0.00	0.00	7.18
Continuity of protective conductors	nr	0.75	7.18	0.00	0.00	7.18
Insulation resistance tests	nr	0.75	7.18	0.00	0.00	7.18
Polarity	nr	0.75	7.18	0.00	0.00	7.18
Earth fault loop impedance	nr	0.75	7.18	0.00	0.00	7.18
Protective devices and earth leakage circuit breaker tests	nr	0.50	4.79	0.00	0.00	4.79
Testing equipment						
multiple earth loop impedance test (weekly hire)	week	0.00	0.00	0.00	0.00	28.00
induction ammeter (digital) (weekly hire)	week	0.00	0.00	0.00	0.00	27.00
portable appliance tester (Megger) (weekly hire)	week	0.00	0.00	0.00	0.00	39.00
RCCB tester (weekly hire)	week	0.00	0.00	0.00	0.00	27.00

289

TESTING AND COMMISSIONING

	Unit	Labour hours	Net labour (£)	Net material (£)	O'heads /profit (£)	Total (£)
Test equipment						
Martindale ring main tester	nr	0.00	0.00	8.70	1.30	10.00
Becker test leads						
combi-check	nr	0.00	0.00	27.74	4.16	31.90
master-check	nr	0.00	0.00	12.85	1.93	14.78
TMK test meters						
digital line earth loop impedance tester	nr	0.00	0.00	175.73	26.36	175.73
Digital ELCB (RCB) triptester		0.00	0.00	143.83	21.57	165.40
Insulation tester	nr	0.00	0.00	148.50	22.27	170.77
Clamp meter SK-8000	nr	0.00	0.00	70.40	10.56	80.96
Clamp meter SK-8100	nr	0.00	0.00	119.90	17.98	137.88
Interchangeable test lead kit	nr	0.00	0.00	9.08	1.36	10.44
Analogue multimeters						
VF-4	nr	0.00	0.00	25.85	3.88	29.73
VF-7	nr	0.00	0.00	39.33	5.90	45.23
TP55N	nr	0.00	0.00	35.15	5.27	40.42
Digital multimeters						
G80	nr	0.00	0.00	32.95	4.94	37.89
G44	nr	0.00	0.00	75.90	11.38	87.28

M & E MEASUREMENT SERVICES

	Unit	Labour hours	Net labour (£)	Net material (£)	O'heads /profit (£)	Total (£)

APPROXIMATE ESTIMATING

Note

The prices contained within this section are guide prices only intended to provide an approximation of costs for total elements of installation. The prices are inclusive of 15% overheads and profits (builders' work excluded)

Prices include PVC cables run behind protective PVC sheath and include all cable, sheath, wiring accessories, but exclude stripping out, builders' work and light fittings.

General wiring

Three bedroom terraced house. 2 two gang 13 amp sockets, 2 way light switches to each room. 4 two gang 13 amp sockets to kitchen plus shaver socket, cooker control unit and extract fan

	Unit	Labour hours	Net labour (£)	Net material (£)	O'heads /profit (£)	Total (£)
	item	0.00	0.00	0.00	0.00	1680.00

Three bedroom semi-detached house. 2 two gang 13 amp sockets, 2 two way light switches to each room. 6 two gang 13 amp sockets to kitchen plus shaver socket, cooker control unit and extract fan

	Unit	Labour hours	Net labour (£)	Net material (£)	O'heads /profit (£)	Total (£)
	item	0.00	0.00	0.00	0.00	1960.00

APPROXIMATE ESTIMATING

	Unit	Labour hours	Net labour (£)	Net material (£)	O'heads /profit (£)	Total (£)
Four bedromm semi-detached house. 7 way SP&N MCB consumer unit, 2 two gang sockets and point in each room with 2 way skirting on the stairs and kitchen. Plus 4 two gang 13A sockets, cooker control unit and extract fan to kitchen	item	0. 00	0. 00	0. 00	0. 00	2000. 00

Chapter 6
Plant and tool hire

The prices contained in this chapter are based upon information supplied by HSS Hire Shops Ltd, 25 Willow Lane, Mitcham, Surrey who have over 160 shops throughout the country (see Yellow Pages). The prices exclude VAT and delivery charges.

	First 24 hrs £	Addit 24 hrs £	Per week £
ACCESS AND SUPPORT			
Narrow tower base, size 1.3 x 1.5m, height			
2.5m	27.50	13.75	55.00
4.5m	40.50	20.25	81.00
6.5m	53.50	26.75	107.00
8.5m	66.50	33.25	133.00
10.5m	79.50	39.75	159.00
Span tower base, size 1.3 x 1.5m or 1.3 x 2.5m, height			
2.5m	27.50	13.75	53.00
4.5m	40.50	20.25	81.00
6.5m	53.50	26.75	107.00
8.5m	66.50	33.25	133.00
10.5m	79.50	39.75	159.00

PLANT AND TOOL HIRE

Access and support (cont'd)		First 24 hrs	Addit 24 hrs	Per week
Alloy ladders				
Double 3.5m extending to 6.2m		9.00	4.50	18.00
5.0m	9.0m	12.50	6.25	25.00
Treble 2.5m	6.0m	9.00	4.50	18.00
3.5m	9.1m	12.50	6.25	25.00
Roof ladders				
Alloy 4.9m, 5.9m and 6.9m		15.00	5.00	25.00
Ladder stay, each		5.40	1.80	9.00
Builders' steps				
8 tread, height 1.5m		7.00	3.50	14.00
10 tread, height 2.1m		8.00	4.00	16.00
12 tread, height 2.7m		9.00	4.50	18.00
Steel trestles				
Nos 1-4: 0.5m extending to 2.4m		-	-	3.20
Scaffold boards				
Length 2.4-3.9m		-	-	1.90
Lightweight staging				
Length : 2.4m		8.00	4.00	10.00
3.0m		9.00	4.50	11.25
3.6m		10.00	5.00	12.50
4.2m		11.00	5.50	13.75
4.8m		13.00	6.50	16.25
6.0m		15.00	7.50	18.75
7.2m		20.00	10.00	25.00

BUILDING AND DECORATING	First 24 hrs £	Addit 24 hrs £	Per week £
Metal locator	9.00	3.00	15.00
Cable avoiding tool	32.40	10.81	54.00
Cat signal generator	19.20	6.40	32.00
Steel props nos 0-4: 1.8m extending to 4.9m (quantity discounts)	-	-	3.20
Jackall prop	9.60	3.20	16.00

Pipe threading

Electric die stock	39.00	45.50	65.00
Ratchet die stock	15.60	5.20	26.00
Pipe-threading machine ½in-4in electric	72.00	24.00	120.00
Pipe-threading machine ½in-2in electric	54.00	18.00	90.00
Pipe pressure tester	15.60	5.20	26.00
Pipe saw	21.60	7.20	36.00

HEATING, COOLING AND DRYING

Industrial heaters - gas

Plaque heater output, 2,500 Btu	10.80	3.60	18.00
Forced air output, 140,000 Btu	28.80	9.60	48.00
Forced air output, 275,000 Btu	39.00	13.00	65.00

Industrial heaters - paraffin

Forced air, 60,000 Btu	27.00	9.00	45.00
Forced air, 85-100,000 Btu	30.00	10.00	50.00
Forced air, 150,000 Btu	39.00	13.00	65.00

Home/office heaters

Cabinet, gas, 3-16,000 Btu	8.40	2.80	14.00
Electric fan heater, 3kW	6.00	2.00	10.00

PLANT AND TOOL HIRE

	First 24 hrs £	Addit 24 hrs £	Per week £
Air conditioning units			
Air conditioner	57.00	19.00	95.00
Space cooler	51.00	17.00	85.00
Cold-air blower	21.00	7.00	35.00
Drying			
Building drier dehumidifier - 300m3	39.00	13.00	65.00
Portable building drier - 140m3	22.80	7.60	38.00
Portable fume extractor	42.00	14.00	70.00
LIGHTING, WELDING AND POWER			
Lighting units			
Gas - large tripod-mounted	12.00	4.00	20.00
Gas - small cylinder-mounted	9.00	3.00	15.00
Electric - tripod-mounted	12.00	4.00	20.00
Plasterers' light - fluorescent	12.00	4.00	20.00
Magnetic 500W flood	11.40	3.80	19.00
Twin 500W flood - 5m tower mast	21.60	7.20	36.00
Festoon lights, industrial (34cm)	12.00	4.00	20.00
Welding			
Arc welder 140/180 amp, 240 volt	19.20	6.40	32.00
MIG welder, 240 volt, medium	25.20	8.40	29.40
No-gas MIG welder	19.20	6.40	32.00
Spot welder, 240 volt (30 amp)	19.20	6.40	32.00
Site welder 20-170 amp, petrol	46.80	15.60	78.00
Welder/generator 300 amp, DC silenced	96.00	32.00	160.00
Oxy/acetylene welding kit	28.80	9.60	48.00

PLANT AND TOOL HIRE

	First 24 hrs £	Addit 24 hrs £	Per week £
Generators (continuous rating)			
3 kVA petrol, 110/240 volt	34.80	11.60	58.00
4 kVA diesel, 110/240 volt	46.80	15.60	78.00
8 kVA diesel, 110/240 volt	72.00	24.00	120.00
15 kVA diesel, 110/240 volt	90.00	30.00	150.00
Transformers			
2.2 kVA	5.70	1.90	9.50
3.0 kVA	8.40	2.80	14.00
5.0 kVA	15.60	5.20	26.00
10 kVA	29.40	9.80	49.00
Extension cable 15m cable/drum	4.50	1.50	7.50
Fourway junction box	7.20	2.40	12.00
Power breaker RCD plug	3.00	1.00	5.00
KANGO BREAKING AND DRILLING			
Hydraulic breakers			
Heavy duty, diesel	57.00	19.00	95.00
Medium duty, petrol	48.00	16.00	80.00
Electric hammers			
Heavy duty breaker (inc. trolley)	33.60	11.20	56.00
Medium duty breaker	18.00	6.00	30.00
Rotary hammers			
Hilti breaker drill, TE 72	18.00	6.00	30.00
Hilti breaker drill, TE 52	18.00	6.00	30.00
Medium-duty breaker drill	16.20	5.40	27.00
Light duty	16.20	5.40	27.00

PLANT AND TOOL HIRE

	First 24 hrs £	Addit 24 hrs £	Per week £
Electric drills			
Two speed percussion drill	8.40	2.80	14.00
Cordless drill	12.00	4.00	20.00
Right angle drill	16.20	5.40	27.00
Four speed drill	16.20	5.40	27.00
Magnetic base drills			
Magnetic drill stand (drill extra)	30.60	10.20	51.00
Magnetic drill stand c/w drill	48.80	15.60	50.60
FIXING, GRINDING AND SANDING			
Fixing tools			
Cartridge hammer	18.00	6.00	30.00
Staple tacker - light duty	6.00	2.00	10.00
Hammer stapler - heavy duty	13.20	4.40	22.00
Nail gun - air operated, automatic	24.00	8.00	40.00
Impact wrench - electric	15.20	5.20	26.00
Screwdriver - electric, auto-feed	16.80	5.60	28.00
Angle grinders			
Angle grinder, 100mm	8.40	2.80	14.00
Angle grinder, 230mm	11.40	3.80	19.00
Angle grinder, 300mm	20.40	6.80	34.00
LIFTING AND MATERIALS HANDLING			
Man lift, height 7.3m	75.00	25.00	125.00
Man lift, height 9.1m	120.00	40.00	200.00
Tirfor winch, TU 16	15.60	5.20	26.00
Tirfor winch, TU 32	19.80	6.60	33.00
Scaffold hoist (200kg)	45.00	15.00	75.00

Chapter 7
General data

The metric system

Linear

1 centimetre (cm)	= 10 millimetres (mm)
1 decimetre (dm)	= 10 centimetres (cm)
1 metre (m)	= 10 decimetres (dm)
1 kilometre (km)	= 1000 metres (m)

Capacity

1 millimetre (ml)	= 1 cubic centimetre (cm^3)
1 centilitre (cl)	= 10 millilitres (ml)
1 decilitre (dl)	= 10 centilitres (cl)
1 litre (l)	= 10 decilitres (dl)

Weight

1 centigram (cg)	= 10 milligrams (mg)
1 decigram (dg)	= 10 centigrams (cg)
1 gram (g)	= 10 decigrams (dg)
1 decagram (dag)	= 10 grams (g)
1 hectogram (hg)	= 10 decagrams (dag)
1 kilogram (kg)	= 10 hectograms (hg) = 1000 grams (g)

Imperial/metric conversions

Linear

1 in	= 25.4mm	1mm = 0.03937in
1 ft	= 304.8mm	1cm = 0.3937in
1 yd	= 914.4mm	1dm = 3.937in
		1m = 39.37in

GENERAL DATA

Square

$1\ in^2 = 645.16mm^2$	$1cm^2 = 0.155\ in^2$
$1\ ft^2 = 0.0929m^2$	$1m^2 = 10.7639\ ft^2$
$1\ yd^2 = 0.8361m^2$	$1m^2 = 1.196\ yd^2$

Cube

$1\ in^3 = 16.3871cm^3$	$1cm^3 = 0.061\ in^3$
$1\ ft^3 = 0.0283m^3$	$1m^3 = 35.3148\ ft^3$
$1\ yd^3 = 0.7646m^3$	$1m^3 = 1.307954\ yd^3$

Capacity

1 fl oz = 28.4ml	1ml = 0.0353 fl oz
1 pt = 0.568 ltr	1dl = 3.52 fl oz
1 gallon = 4.546 ltr	1 ltr = 1.7598 pt

Weight

1 oz = 28.35g	1g = 0.035 oz
1 lb = 0.4536kg	1kg = 35.274 oz
1 st = 6.35kg	1t = 2204.6 lb
1 ton = 1.016t	1t = 0.9842 ton

Temperature equivalents

In order to convert Fahrenheit to Celsius deduct 32 and multiply by 5/9. To convert Celsius to Fahrenheit multiply by 9/5 and add 32.

Fahrenheit		Celsius
230		110.0
220		104.4
212	Boiling point	100.0
210		98.9
200		93.3
190		87.8
180		82.2
170		76.7
160		71.1
150		65.6
140		60.0

GENERAL DATA

Fahrenheit		Celsius
130		54.4
120		48.9
110		43.3
90		32.2
80		26.7
70		21.1
60		15.6
50		10.0
40		4.4
32	Freezing point	0.0
30		-1.1
20		-6.7
10		-12.2
0		-17.8

Melting points of materials

Aluminium	$658^{\circ}C$
Brass	$927\text{-}1010^{\circ}C$
Bronze	$912^{\circ}C$
Cast iron	$1186^{\circ}C$
Copper	$1083^{\circ}C$
Lead	$327^{\circ}C$
Nickel	$1452^{\circ}C$
Steel	$1371^{\circ}C$
Tin	$230^{\circ}C$
Zinc	$419^{\circ}C$

Milled lead to BS1178

Code	Thickness	Weight	Colour code
3	1.32mm	14.97 kg/m^2	Green
4	1.80mm	20.41 kg/m^2	Blue
5	2.24mm	25.40 kg/m^2	Red
6	2.65mm	30.05 kg/m^2	Black
7	3.15mm	35.72 kg/m^2	White
8	3.55mm	40.26 kg/m^2	Orange

Standard wire gauge and metric equivalent

SWG	mm	SWG	mm
3	6.40	15	1.83
4	5.89	16	1.63
5	5.38	17	1.42
6	4.88	18	1.21
7	4.47	19	1.02
8	4.06	20	0.91
9	3.63	21	0.81
10	3.25	22	0.71
11	2.95	23	0.61
12	2.65	24	0.56
13	2.34	25	0.51
14	2.03	26	0.46

Alternating current data

$$\text{Power factor} = \frac{\text{watts}}{\text{volts x amperes}}$$

Watts in single-phase circuit = volts x amperes x PF
Watts in three-phase circuit = $\sqrt{3}$ x volts x amperes x PF
Star windings: Line voltage = 1.732 x phase voltage
Mesh windings: Line voltage = phase voltage
Speed of AC motors in rev/min

$$\text{Synchronous} = \frac{\text{Frequency x 60}}{\text{Number of pairs of poles}}$$

$$\text{Induction} = \frac{\text{Frequency x 60}}{\text{Number of pairs of poles}} \text{ x (1 - pu slip)}$$

302

GENERAL DATA

Standard electric motors

Average efficiencies and power factors

| kW | DC | | Single | | | Two | | Three | |
| | | Split | phase | Capacitor | | phase | | phase | |
	Eff.	Eff.	PF	Eff.	PF	Eff.	PF	Eff.	PF
0.75	0.76	0.68	0.74	0.74	0.94	0.73	0.79	0.76	0.83
4.0	0.83	0.78	0.83	0.82	0.92	0.84	0.84	0.82	0.86
7.5	0.86	0.81	0.84	0.84	0.93	0.87	0.86	0.86	0.88
15	0.88	0.83	0.85	0.86	0.94	0.88	0.88	0.89	0.89
37	0.90	0.85	0.87	0.88	0.94	0.91	0.90	0.91	0.91
55 and larger	0.92	0.86	0.87	0.89	0.95	0.91	0.90	0.91	0.91

Formulae:

Current taken by direct current motor

$$I = \frac{1000P}{E \times V}$$

I = Current taken by motor
P = Output of motor in kilowatts
E = Efficiency expressed as decimal
V = Voltage

Current taken by induction motors

$$I = \frac{1000P}{E \times m \times V_{PF} \times PF}$$

For 3-phase supply V_{ph} = Line voltage divided by $\sqrt{3}$
I = Current taken by motor
P = Output of motor in kilowatts
E = Efficiency expressed as decimal
m = No. of phases
V_{ph} = Voltage per phase
PF = Power factor

303

GENERAL DATA

Approximate current in amperes at full load taken by single-phase induction motors at various voltages assuming average power factor and efficiency.

	Single-phase voltage					
	Split Phase Voltage			Capacitor	Voltage	
kW	240	415	480	240	415	480
0.75	6.2	3.6	3.1	4.5	2.6	2.3
1.5	10.8	6.3	5.4	8.8	5.1	4.4
2.2	15.4	8.9	7.7	13.0	7.5	6.5
3.0	20.0	11.5	10.0	16.9	9.8	8.5
4.0	25.5	15.0	13.0	22.0	12.71	1.0
5.5	34.0	20.0	17.0	30.0	17.51	5.0
7.5	46	26	23	40	23	20
11	67	39	33	58	34	29
15	88	51	44	77	44	38
18.5	110	64	55	97	56	48
22	130	75	65	114	66	57
30	173	100	87	150	87	75
37	210	122	105	188	109	94
45	252	146	126	225	130	112
55	312	180	156	276	160	138
75	410	237	205	364	210	182

Approximate current in amperes per phase at full load taken by 2- and 3-phase induction motors at various voltages assuming average power factor and efficiency.

Two-phase voltage			Three-phase voltage	
kW	200	400	346	415
0.75	2.9	1.4	1.4	2.0
1.5	5.7	2.9	2.9	3.5
2.2	8.4	4.2	6.0	5.0

Two-phase voltage			Three-phase voltage	
kW	200	400	346	415
3.0	11.0	5.6	7.8	6.5
4.0	15.0	7.3	9.6	8.0
5.5	20	11	13	11
7.5	26	13	17	14
11	38	19	25	21
15	50	25	33	28
18.5	62	31	42	35
22	74	37	48	40
30	96	48	66	55
37	118	59	79	66
45	140	70	96	80
55	171	86	120	100
75	228	114	162	135
112	336	168	240	200
150	446	223	310	260
190	564	282	390	325
224	672	336	460	385

For mid-wire of two-phase three-wire, multiply two-phase current by 1.41.

Approximate current in amperes at full load at various voltages, assuming average efficiency.

	Voltage		Voltage	
kW	110	240	480	600
0.75	10.0	4.5	2.5	2.0
1.5	18.0	8.0	4.0	3.5
2.2	26.0	11.0	6.0	5.0

Voltage			Voltage	
kW 110		240	480	600
4.0	45.0	20.5	9.5	8.0
5.5	61.0	28.0	13.0	11.0
7.5	81.0	36.0	18.0	15.0
11	115.0	54.0	27.0	21.0
15	155.0	71.0	35.0	28.0
22	225.0	105.0	53.0	42.0
30	300.0	135.0	69.0	55.0
37	375.0	170.0	85.0	68.0
55	560.0	258.0	128.0	103.0
75	740.0	340.0	170.0	135.0
95	900.0	410.0	205.0	170.0
112	-	505.0	250.0	200.0

Kilowatts into horsepower

	0	1	2	3	4
kW	hp	hp	hp	hp	hp
0	-	1.341	2.682	4.023	5.364
10	13.410	14.751	16.092	17.433	18.774
20	26.820	28.161	29.502	30.843	32.184
30	40.231	41.572	42.913	44.254	45.595
40	53.641	54.982	56.323	57.664	59.005
50	67.051	68.392	69.733	71.074	72.415
60	80.461	81.802	83.143	84.484	85.825
70	93.871	95.212	96.553	97.894	99.235
80	107.28	108.62	109.96	111.30	112.65
90	120.69	122.03	123.37	124.71	126.06
100	134.10	135.44	136.78	138.12	139.47

Kilowatts into horsepower (cont'd)

kW	5 hp	6 hp	7 hp	8 hp	9 hp
0	6.705	8.046	9.387	10.728	12.069
10	20.115	21.456	22.797	24.138	25.479
20	33.525	34.866	36.208	37.549	38.890
30	46.936	48.277	49.618	50.959	52.300
40	60.346	61.687	63.028	64.369	65.710
50	73.756	75.097	76.438	77.779	79.120
60	87.166	88.507	89.848	91.189	92.530
70	100.58	101.92	103.26	104.60	105.94
80	113.99	115.33	116.67	118.01	119.35
90	127.40	128.74	130.08	131.42	132.76
100	140.81	142.15	143.49	144.83	146.17

Horsepower into kilowatts

hp	0 kW	1 kW	2 kW	3 kW	4 kW
0	0	0.746	1.491	2.237	2.983
10	7.457	8.203	8.948	9.694	10.440
20	14.914	15.660	16.405	17.151	17.897
30	22.371	23.117	23.862	24.608	25.354
40	29.828	30.574	31.319	32.065	32.811
50	37.285	38.031	38.776	39.522	40.268
60	44.742	45.488	46.233	46.979	47.725
70	52.199	52.945	53.691	54.436	55.182
80	59.656	60.402	61.148	61.893	62.639
90	67.113	67.859	68.605	69.350	70.096
100	74.570	75.316	76.062	76.807	77.553

hp	5 kW	6 kW	7 kW	8 kW	9 kW
0	3.729	4.474	5.220	5.966	6.711
10	11.186	11.031	12.677	13.423	14.168
20	18.643	19.388	20.134	20.880	21.625
30	26.100	26.845	27.591	28.337	29.082
40	33.557	34.302	35.048	35.794	36.539
50	41.014	41.759	42.505	43.251	43.996
60	48.471	49.216	49.962	50.708	51.453
70	55.928	56.673	57.419	58.165	58.910

GENERAL DATA

Horsepower into kilowatts (cont'd)

80	63.385	64.130	64.876	65.622	66.367
90	70.842	71.587	72.333	73.079	73.824
100	78.299	79.044	79.790	80.536	81.281

Electrical British Standards in common use

BS	Description
77	Voltages for AC transmission and distribution systems.
88	Cartridge fuses for voltages up to and including 1000V AC and 1500V DC (parts 1, 2, 4 & 5).
89	Direct-acting electrical indicating instruments.
142	Electrical protective relays.
148	Insulating oil (low viscosity) for transformers and switchgear.
159	Busbars and busbar connections.
162	Electric power switchgear and associated apparatus
171	Power transformers (Parts 1, 2, 4 and 5).
775	AC contactors for voltages above 1kV and up to and including 12kV (Part 2).
822	Terminal markings for electrical machinery and apparatus (Part 6).
1363	13A plugs, switched and unswitched socket-outlets and boxes.
2692	Fuses for voltages over 1000V AC (Parts 1 and 2).
3676	Switches for domestic and similar purposes (for fixed or portable mounting).
3938	Current transformers.
3941	Voltage transformers.
4177	Cooker control units rated at 30A and 45A 250V single-phase only.
4417	Semiconductor rectifier equipments.
4568	Steel conduit and fittings with metric threads of ISO form for electrical installations (Parts 1 and 2).
4727	Glossary of electrotechnical power, telecommunications, electronics, lighting, and colour terms.
4752	Switchgear and control gear for voltages up to and including 1000V AC and 1200V DC.
4999	General requirements for rotating electrics machines.
5000	General purpose induction motors (Part 10). Small power electric motors and generators (Part 11). Machines for miscellaneous applications (Part 99).
5227	AC metal-enclosed switchgear and control gear of rated voltages above 1kV and up to and including 72.5kV.
5311	AC circuit breakers of rated voltage above 1kV (Parts 1-7).
5405	Codes of practice for the maintenance of electrical switchgear for voltages up to and including 145kV.

5419 Airbreak switches, airbreak disconnectors, airbreak switch disconnectors and fuse combination units for voltages up to and including 1000V AC and 1200V DC.

5424 Control gear for voltages up to and including 1000V AC and 1200V DC.

5463 AC switches of rated voltage above 1kV.

5685 Electricity meters (Parts 1-5).

6004 PVC-insulated cables (non-armoured) for electric power and lighting.

6007 Rubber-insulated cables for electric power and lighting.

6480 Impregnated paper-insulated cables for electricity supply.

6485 PVC-covered conductors for overhead power lines.

6500 Insulated flexible cords.

Codes of practice

CP 1011 Maintenance of electric motor control gear.

CP 1013 Earthing.

CP 1015 Electrical equipment of industrial machines.

CP 1017 Distribution of electricity on construction and building sites.

Statutory regulations covering electrical installations

Type of premises and/or equipment covered:	Legislation title:
General installations with exemptions	Electricity Supply Regulations 1937 currently under review
Building generally with exemptions (Scotland only)	Building Standards (Scotland) Regulations 1981
Factory installations	Electricity (Factories Act) Special Regulations 1908 and 1944
Cinematograph installations	Cinematographic Act 1909 and/or Cinematographic Act 1952
Coal mines including certain other mines	Coal and Other Mines (Electricity) Regulations 1956

Quarry installations

Quarries (Electricity)
Regulations 1956

Metalliferous mines

Miscellaneous Mines (Electricity)
Regulations 1956

Agricultural and
horticultural installations

Agriculture (Stationary Machinery)
Regulations 1959

Electrical equipment

Electrical Equipment (Safety)
Regulations 1975 - (under review
draft E.E. (Safety) Regulations
1987 out for comment)

Accessories

Plugs and Sockets etc. (Safety)
Regulations 1987

INDEX